"互联网+" 创新创业实践系列教材

鸿蒙应用开发教程

钟元生　林生佑　主编
李浩轩　吴　冕　副主编

清华大学出版社
北京

内容简介

本书系统讲解鸿蒙应用开发的基础知识,既有基本语法与基本应用,又有案例分析,使读者能理论联系实际,寓教于练、寓教于用,是鸿蒙编程的快速入门书籍。

本书分为七章,内容包括 HarmonyOS 简介与环境搭配、HarmonyOS 界面编程基础、HarmonyOS 事件处理、Ability、数据管理、公共事件、通知与日志、综合案例——"远程闹钟"。

本书内容充实、资料新颖、案例丰富、条理清晰,可作为软件工程、计算机科学与技术等专业本科生和研究生的教材,也可作为有志于开发基于鸿蒙应用程序的读者的参考书。

本书封面贴有清华大学出版社防伪标签,无标签者不得销售。
版权所有,侵权必究。举报: 010-62782989, beiqinquan@tup.tsinghua.edu.cn。

图书在版编目(CIP)数据

鸿蒙应用开发教程/钟元生,林生佑主编. —北京:清华大学出版社,2022.8(2025.1重印)
"互联网+"创新创业实践系列教材
ISBN 978-7-302-61056-4

Ⅰ.①鸿⋯ Ⅱ.①钟⋯ ②林⋯ Ⅲ.①移动终端-应用程序-程序设计-高等学校-教材 Ⅳ.①TN929.53

中国版本图书馆 CIP 数据核字(2022)第 096440 号

责任编辑:袁勤勇　常建丽
封面设计:杨玉兰
责任校对:焦丽丽
责任印制:杨　艳

出版发行:清华大学出版社
网　　址: https://www.tup.com.cn, https://www.wqxuetang.com
地　　址: 北京清华大学学研大厦 A 座　　邮　编: 100084
社 总 机: 010-83470000　　邮　购: 010-62786544
投稿与读者服务: 010-62776969, c-service@tup.tsinghua.edu.cn
质量反馈: 010-62772015, zhiliang@tup.tsinghua.edu.cn
课件下载: https://www.tup.com.cn, 010-83470236

印 装 者: 三河市铭诚印务有限公司
经　　销: 全国新华书店
开　　本: 185mm×260mm　　印　张: 20.75　　字　数: 505 千字
版　　次: 2022 年 8 月第 1 版　　印　次: 2025 年 1 月第 5 次印刷
定　　价: 68.00 元

产品编号: 095742-01

前 言

鸿蒙操作系统是一款连接万物的操作系统,为我国自主研发的新型操作系统。它通过分布式架构实现了设备互联互通,能够自适应不同的设备和使用场景。鸿蒙内核性能更高、功耗更低、用户体验更稳定,应用和服务更加智能化。至 2023 年 6 月,它已经拥有 38 万个开发者、4.5 万个应用和和 8 亿用户,鸿蒙生态发展迅速,是我国对世界信息技术的一大贡献。

在学习鸿蒙程序设计时,宜结合党的二十大关于"教育、科技、人才"三位一体的精神,引导学生领悟"高水平科技自立自强"对我国发展的极端重要性,理解践行"新型举国体制"的重要作用,引导学生抓住鸿蒙编程学习的大好机会,积极投身于基于鸿蒙的应用程序开发工作,致力打造基于国产操作系统的软件生态系统,建立数据安全观,破解西方国家对我国的"卡脖子"战略。同时,还要结合鸿蒙编程技术学习,引导学生理解高水平改革开放、积极构建国内国际双循环、推动世界各国合作共赢的价值。

现在搭载鸿蒙操作系统的终端越来越普及。谁更早掌握鸿蒙操作系统开发技术,谁就占有发展先机。许多学生都想快速学习鸿蒙系统编程,大家都希望有一本好的教材。为此,我们将以往编写《Android 编程》等 App 系统教材的成功经验移植到《鸿蒙应用开发教程》教材中,通过近一年的案例学习、教学设计和应用交流,完成本书。本书努力做到:

(1) 既介绍鸿蒙开发基本语法、基本知识和基础应用,又介绍可以直接运行的应用教学案例,使教师容易教学,使学生能寓教于练、寓教于用。

(2) 不仅讲解注重语法细节,而且循序渐进地引导和启发学生构建自己的知识体系,包括用图解法详细分析鸿蒙应用程序的结构、运行过程以及各部分之间的调用关系,演示鸿蒙应用的开发流程。

(3) 重点关注手机应用中的常见案例,将有关知识串联起来。结合学生使用鸿蒙系统手机的体验,逐步引导学生深入思考其内部实现。每章后的习题,可帮助学生自测。

本书假定读者懂一些基本的 Java 语法知识,具有一定的 Java 编程经验。如果没有 Java 基础,在阅读本书遇到 Java 知识时,建议补充学习一些相关知识。

书中示例较多,源代码较长。本书注重示例的程序分析,为了方便介绍知识点,压缩篇幅,仅列出一些关键代码,读者可以从相应网站下载完整源代码并直接运行。

为方便阅读,书中的每段代码都引入了代码编号,部分关键语句加了注释并给出程序在资源包中的位置,样例如下所示。

程序清单4-25：hmos\ch04\01\AbilityTest\entry\src\main\java\com\
example\abilitytest\slice\MainAbilitySlice.java

```
1   @Override
2   public void onStart(Intent intent) {
3       super.onStart(intent);
4       //加载 XML 布局
5       super.setUIContent(ResourceTable.Layout_ability_main);
6       ...
7       //启动本地设备 Service
8       startupLocalService();
9       ...
10  }
```

其中，左边的1,2,3,…表示行号，右边的"super.onStart(intent);"才是真实的程序代码内容。第6、9行的"…"表示省略了与此处示例功能无关的代码。在关键代码处也通过注释的方式加以说明。

为了方便读者学习、交流与共享资源，我们提供了本书相关资源，请到清华大学出版社官网下载。

本书由钟元生、林生佑担任主编，由李浩轩、吴冕担任副主编。各章分工如下：钟元生负责第1章并参与其余各章修改，吴冕、钟元生负责第2章和第3章，李浩轩、钟元生负责第4章和第5章，林生佑负责第6章和第7章。

在此，特别感谢支持本书编写出版工作的江西财经大学软件与物联网工程学院、清华大学出版社有关领导和编辑的帮助。

希望本书的出版有助于"鸿蒙应用开发"任课老师更快地教好鸿蒙编程课，也能帮助使用本书的学生更快、更扎实地掌握鸿蒙应用开发技能。书中难免存在不当之处，还望大家批评指正，以便再版时完善。

编　者

于江西财经大学麦庐园

2023年9月

鸿蒙应用开发教程的思政元素与教学实践

目 录

第 1 章　HarmonyOS 简介与环境搭建　<<< 1

  1.1　初识 HarmonyOS ……………………………………………………… 2
    1.1.1　HarmonyOS 概述 ……………………………………………… 2
    1.1.2　HarmonyOS 的体系结构 ……………………………………… 2
  1.2　搭建 HarmonyOS 开发环境 …………………………………………… 3
  1.3　HarmonyOS 项目运行过程分析 ……………………………………… 18
    1.3.1　HarmonyOS 应用程序结构分析 ……………………………… 18
    1.3.2　HarmonyOS 应用程序运行过程分析 ………………………… 19
  1.4　本章小结 ………………………………………………………………… 23
  1.5　课后习题 ………………………………………………………………… 23

第 2 章　HarmonyOS 界面编程基础　<<< 24

  2.1　HarmonyOS 界面设计概述 …………………………………………… 26
  2.2　HarmonyOS 基础界面控件 …………………………………………… 26
    2.2.1　Text 组件 ……………………………………………………… 26
    2.2.2　Button 组件 …………………………………………………… 31
    2.2.3　TextField 组件 ………………………………………………… 34
    2.2.4　TabList 组件 …………………………………………………… 35
    2.2.5　DatePicker 组件 ……………………………………………… 41
    2.2.6　TimePicker 组件 ……………………………………………… 45
    2.2.7　Checkbox 组件 ………………………………………………… 48
    2.2.8　Image 组件 …………………………………………………… 56
  2.3　HarmonyOS 布局管理器 ……………………………………………… 57
    2.3.1　DirectionalLayout 布局 ……………………………………… 57
    2.3.2　DependentLayout 布局 ……………………………………… 62
    2.3.3　StackLayout 布局 …………………………………………… 64
    2.3.4　TableLayout 布局 …………………………………………… 65
    2.3.5　PositionLayout 布局 ………………………………………… 67
    2.3.6　AdaptiveBoxLayout 布局 …………………………………… 69
  2.4　HarmonyOS 高级界面控件 …………………………………………… 74
    2.4.1　ListContainer 列表 …………………………………………… 74
    2.4.2　CommonDialog 对话框 ……………………………………… 78

　　　　2.4.3　RadioContainer 单选按钮容器 ……………………………… 79
　　2.5　本章小结 ………………………………………………………………… 82
　　2.6　课后习题 ………………………………………………………………… 83

第 3 章　HarmonyOS 事件处理　<<<< 84
　　3.1　HarmonyOS 基于监听的事件处理 ……………………………………… 86
　　3.2　HarmonyOS 线程管理 …………………………………………………… 90
　　　　3.2.1　线程管理接口说明 …………………………………………… 90
　　　　3.2.2　线程管理开发步骤 …………………………………………… 92
　　3.3　HarmonyOS 线程间通信 ………………………………………………… 100
　　　　3.3.1　线程间通信场景介绍 ………………………………………… 100
　　　　3.3.2　线程间通信接口介绍 ………………………………………… 102
　　3.4　本章小结 ………………………………………………………………… 113
　　3.5　课后习题 ………………………………………………………………… 114

第 4 章　Ability 与 Intent　<<<< 115
　　4.1　Ability 介绍 ……………………………………………………………… 116
　　　　4.1.1　创建一个 Ability ……………………………………………… 117
　　　　4.1.2　Ability 的配置 ………………………………………………… 117
　　4.2　Page Ability ……………………………………………………………… 118
　　　　4.2.1　Page Ability 与 AbilitySlice …………………………………… 118
　　　　4.2.2　创建 Page Ability ……………………………………………… 118
　　　　4.2.3　AbilitySlice 路由配置 ………………………………………… 120
　　　　4.2.4　Page Ability 生命周期 ………………………………………… 121
　　　　4.2.5　AbilitySlice 生命周期 ………………………………………… 123
　　　　4.2.6　AbilitySlice 间的导航 ………………………………………… 125
　　　　4.2.7　不同 Page Ability 间的导航 …………………………………… 126
　　　　4.2.8　用户注册案例 ………………………………………………… 129
　　4.3　Service Ability …………………………………………………………… 146
　　　　4.3.1　创建 Service Ability …………………………………………… 146
　　　　4.3.2　启动 Service Ability …………………………………………… 150
　　　　4.3.3　停止 Service Ability …………………………………………… 154
　　　　4.3.4　连接 Service Ability …………………………………………… 156
　　　　4.3.5　断开 Service Ability …………………………………………… 159
　　　　4.3.6　利用 Service Ability 处理数据 ………………………………… 160
　　　　4.3.7　前台 Service Ability …………………………………………… 164
　　　　4.3.8　示例的完整代码 ……………………………………………… 165
　　4.4　Data Ability ……………………………………………………………… 172

| 4.4.1 URI 介绍 ································· 172
| 4.4.2 URI 示例 ································· 173
| 4.4.3 创建 Data Ability ···················· 173
| 4.4.4 Data Ability 相关类 ················ 175
| 4.5 本章小结 ·· 177
| 4.6 课后习题 ·· 177

第 5 章　数据管理　<<< 178
| 5.1 关系数据库 ·· 179
| 5.1.1 关系数据库介绍 ························ 179
| 5.1.2 约束与限制 ······························· 180
| 5.1.3 关系数据库相关类 ···················· 180
| 5.1.4 关系数据库开发步骤 ················ 182
| 5.1.5 基于 Data Ability 的关系数据库操作案例 ······ 184
| 5.2 分布式数据服务 ·································· 201
| 5.2.1 分布式数据服务介绍 ················ 201
| 5.2.2 单版本分布式数据库 ················ 201
| 5.2.3 分布式数据服务相关类 ············ 202
| 5.2.4 单版本分布式数据服务案例 ····· 203
| 5.3 本章小结 ·· 218
| 5.4 课后习题 ·· 218

第 6 章　公共事件、通知与日志　<<< 219
| 6.1 公共事件 ·· 220
| 6.1.1 四种公共事件 ··························· 221
| 6.1.2 公共事件相关类 ······················· 221
| 6.1.3 无序的公共事件开发 ················ 224
| 6.1.4 带权限的公共事件开发 ············ 227
| 6.1.5 有序的公共事件开发 ················ 229
| 6.1.6 黏性的公共事件开发 ················ 231
| 6.2 通知 ·· 233
| 6.2.1 通知相关类 ······························· 233
| 6.2.2 通知开发示例 ··························· 235
| 6.2.3 单击通知栏事件 ······················· 237
| 6.3 日志 ·· 240
| 6.4 本章小结 ·· 241
| 6.5 课后习题 ·· 241

第7章 综合案例——"远程闹钟" <<< 242

7.1 "远程闹钟"概述 ………………………………………………… 244
7.2 Spring Boot 服务器端设计 …………………………………… 245
  7.2.1 Spring Boot 技术简介 ………………………………… 245
  7.2.2 Spring Boot 项目开发环境 …………………………… 245
  7.2.3 数据库设计 ……………………………………………… 247
  7.2.4 "远程闹钟"服务器搭建 ………………………………… 250
  7.2.5 部署服务器 ……………………………………………… 260
7.3 "远程闹钟"手机端应用设计 …………………………………… 261
  7.3.1 闹钟显示模块 …………………………………………… 262
  7.3.2 添加闹钟模块 …………………………………………… 276
  7.3.3 删除闹钟模块 …………………………………………… 281
7.4 "远程闹钟"手表端应用设计 …………………………………… 292
  7.4.1 手表端应用的创建 ……………………………………… 292
  7.4.2 闹铃定时播放模块 ……………………………………… 295
7.5 本章小结 ………………………………………………………… 311
7.6 课后习题 ………………………………………………………… 311

附录 A SQL 语句使用简介 <<< 312

 A.1 SQL 介绍 ………………………………………………… 312
 A.2 SQL 项目表设计 ………………………………………… 312
 A.3 创建 SQL 表 ……………………………………………… 313
 A.4 SQL 的查询、增加、修改、删除操作方法 ……………… 318
 A.5 小结 ……………………………………………………… 322

第 1 章

HarmonyOS 简介与环境搭建

课程思政 1

本章要点

- 初识 HarmonyOS
- 搭建 HarmonyOS 开发环境
- HarmonyOS 项目运行过程分析

本章知识结构图（见图 1-1）

图 1-1　本章知识结构图

本章示例

HarmonyOS 环境搭建成功后，创建第一个 HarmonyOS 项目，启动远程模拟器，程序效果如图 1-2 所示。

本章是 HarmonyOS 应用程序开发的准备章节，主要介绍什么是 HarmonyOS，HarmonyOS 的发展历程，如何搭建 HarmonyOS 开发环境，然后通过一个简单的 HelloWorld 程序讲解 HarmonyOS 项目的创建、运行过程以及 HarmonyOS 应用程序目录结构中各文件的作用及关系等。本章是学好 HarmonyOS 的基础，是学习其他章节前必须掌握的内容。

图 1-2　本章项目效果图

1.1 初识 HarmonyOS

在 5G 逐渐普及、万物互联的数字化新时代,移动互联网行业经历了飞速的发展,数字化新时代的到来需要新的操作系统,HarmonyOS 也由此应运而生,它的诞生表明,未来科技领域正从智能手机转移到智能穿戴设备和各种智能 IoT(物联网)设备,新的变革已经到来。那么,我们为什么选择 HarmonyOS?它究竟有什么特点?下面从不同的角度来认识 HarmonyOS。

1.1.1 HarmonyOS 概述

HarmonyOS 的中文名称是"鸿蒙",有着浓厚的中国文化气息。Harmony 中文翻译为"和谐",华为希望这个系统可以为世界带来更多的和谐与便利。鸿蒙的发布,在世界科技强国持续打压中国的局势下,更加具有里程碑意义。

随着 5G 物联网时代的到来,鸿蒙是面向未来物联网时代的新系统,更是一款面向全场景的分布式智慧操作系统,致力于创造一个万物互联的智能世界。该系统能够支持多种终端设备,将用户在生活实际应用中接触的各种智能终端实现极速发现、极速连接、硬件互助、资源共享,用合适的设备提供统一、便利、安全的智慧化全场景体验,HarmonyOS 可从根本上帮助消费者解决智能终端体验割裂的问题。

2019 年,HarmonyOS 1.0 版本正式亮相,2020 年 9 月 10 日在华为开发者大会上发布 HarmonyOS 2.0,从此鸿蒙操作系统的神秘面纱终于被揭开,HarmonyOS 2.0 给应用带来更多的流量入口,给设备带来更好的互联体验。该系统实现了硬件互助、资源共享,同时也能够实现一次开发,多端部署以及统一 OS,弹性部署,并且该系统在安全和隐私方面也有较大的提升。2021 年年底,搭载鸿蒙操作系统的设备数量达 3 亿台,其中华为设备超过 2 亿台,面向第三方合作伙伴的各类终端设备数量超过 1 亿台。2021 年 9 月 10 日起,HarmonyOS 面向智慧屏、可穿戴设备、车机等 RAM 在 128KB~128MB 终端设备开放源代码,2021 年 10 月以后,面向 4GB 以上所有设备开源,HaromnyOS 的稳定版本定于 2022 年 9 月发布,具有改进的 UI(用户界面)和更好的整体性能,目前在数百万开发者的支持下,HarmonyOS 已经成为全球第三大移动应用生态。

1.1.2 HarmonyOS 的体系结构

HarmonyOS 整体遵守分层的层次化设计,一共 4 层,从下向上依次为:内核层、系统服务层、框架层和应用层。系统功能按照"系统>子系统>功能/模块"逐级展开,在多设备部署场景下,支持根据实际需求裁剪某些非必要的子系统或功能/模块,如图 1-3 所示。

下面分别对各个层次的功能做简要介绍。

1. 内核层

内核子系统:最底层的内核主要由 Linux Kernel 和 LiteOS 构成,负责操作系统的基本功能,比如线程的调度与内存管理。Linux 是一种开源免费操作系统内核,华为基于 Linux 来开发鸿蒙系统,而安卓系统也基于 Linux 开发,不但解决了生态问题,又可以很

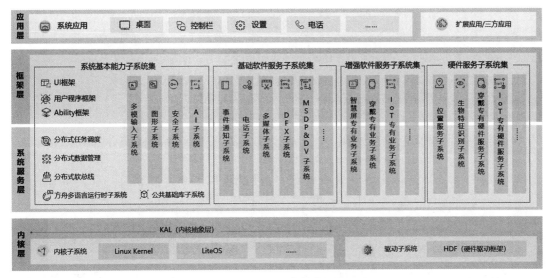

图 1-3　HarmonyOS 技术架构图

好地兼容现有的安卓 App。LiteOS 内核主要是华为针对物联网设备发布的一个轻量级操作系统,以轻量级、低功耗、快速启动、安全等为特性。

驱动子系统:鸿蒙的硬件抽象框架叫作 HDF,提供统一的外设访问能力和驱动开发管理框架。它是鸿蒙 OS 硬件生态开发的基础。

2. 系统服务层

系统服务层为应用程序的运行提供各类服务,比如多设备的调度、定位、生物识别等。它是 HarmonyOS 的核心能力集合,涵盖了 4 个子系统集,分别是系统基本能力子系统集、基础软件服务子系统集、增强软件服务子系统集、硬件服务子系统集。

3. 框架层

框架层为 HarmonyOS 应用开发提供了 Java/C/C++/JavaScript 等多语言的用户程序框架和 Ability 框架,以及各种软硬件服务对外开放的多语言框架 API;同时为采用 HarmonyOS 的设备提供了 C/C++/JavaScript 等多语言的框架 API,开发者可以直接调用 API 构建自己的应用程序。

4. 应用层

最顶上的一层叫应用层。HarmonyOS 的应用由一个或多个 FA(Feature Ability)或 PA(Particle Ability)组成。FA 和 PA 是 HarmonyOS 应用的基本组成单元,能够实现特定的业务功能。其中,FA 有 UI(用户界面),提供与用户交互的能力,而 PA 无 UI,提供后台运行任务的能力以及统一的数据访问抽象。应用层直接与用户打交道,我们使用的浏览器、备忘录等软件,都属于应用层。

1.2　搭建 HarmonyOS 开发环境

搭建 HarmonyOS 开发环境,需要 4 个步骤,如表 1-1 所示。

表 1-1　HarmonyOS 开发环境配置表

步骤	操作步骤	操作指导	说明
1	华为开发者认证	登录官网进行认证	只有认证后才可以获得更多开发权益
2	软件安装	登录官网进行软件安装	安装 DevEco Studio
3	配置开发环境	下载 HarmonyOS SDK	如果网络不能直接访问 Internet，请参照华为官网配置代理
4	运行 HelloWorld	创建工程	运行 Demo 工程，验证环境是否已经配置完成

下载之前需要进行华为开发者认证，才可以获得更多开发权益。打开 https://developer.harmonyos.com，单击右上角的"注册"按钮，如图 1-4 所示。

图 1-4　HarmonyOS 应用开发官网

填写手机号与短信验证码，输入合法格式的密码后即可单击"注册"按钮，如图 1-5 所示。

图 1-5　华为账户注册

返回首页后，可以发现账号已经登录，将鼠标移至账号名处，会显示一个子页面，单击

"实名认证",进行下一步,如图1-6所示。

图1-6 准备实名认证

这时已成功注册华为账号,只需完成实名认证,即可成为华为开发者联盟合作伙伴,获得更多开发、分发等服务权益。单击个人开发者下方的"下一步"按钮,如图1-7所示。

图1-7 开发者实名认证选择

根据项目需求判断是否有以下敏感应用上架到应用市场,建议选"是",便于日后应用上架,然后单击"下一步"按钮,如图1-8所示。

接下来进行个人银行卡实名认证,输入真实姓名、银行卡号等信息,单击"下一步"按钮,补齐资料信息后即可快速完成认证,接下来就可以下载了,如图1-9所示。

然后需要下载鸿蒙应用程序开发工具,DevEco Studio是以IntelliJ IDEA Community开源版本为基础,针对鸿蒙应用程序开发框架而设计的IDE(集成开发环境),是面向全场景多设备的一站式分布式应用开发平台。DevEco Studio启动界面如图1-10所示。

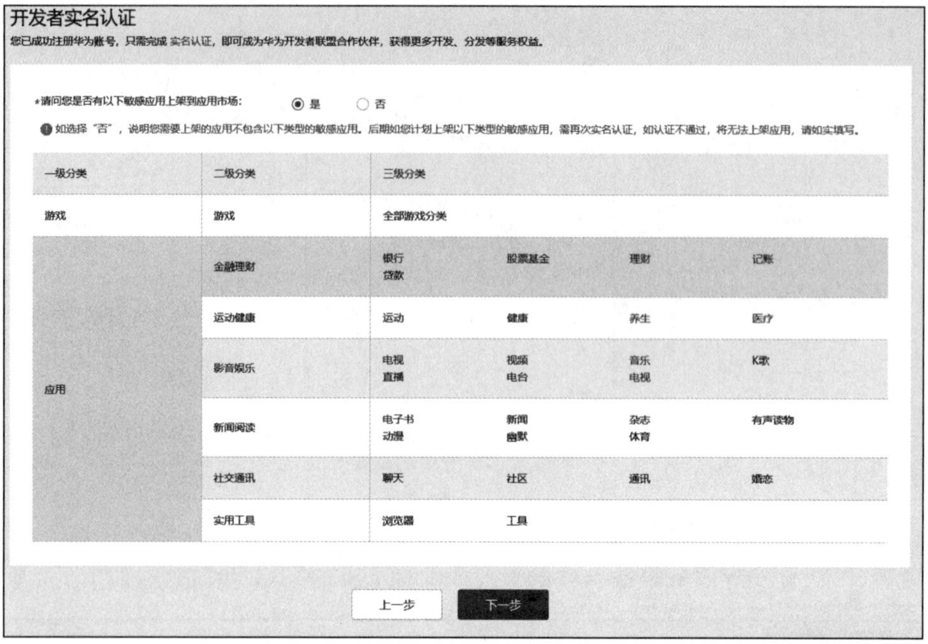

图 1-8 应用分类确认

图 1-9 个人银行卡实名认证

下载地址为 https://developer.harmonyos.com/cn/develop/deveco-studio，根据操作系统选择相应的版本，我们选择 Windows（64-bit）位版本，如图 1-11 所示。

解压 zip 安装包后得到 exe 格式的安装文件，双击打开该文件进行安装，安装路径全英文，至少需要 1.9GB 的磁盘空间，如图 1-12 所示。

第1章　HarmonyOS 简介与环境搭建

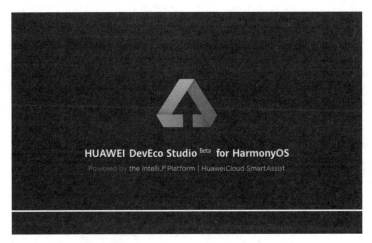

图 1-10　DevEco Studio 启动界面

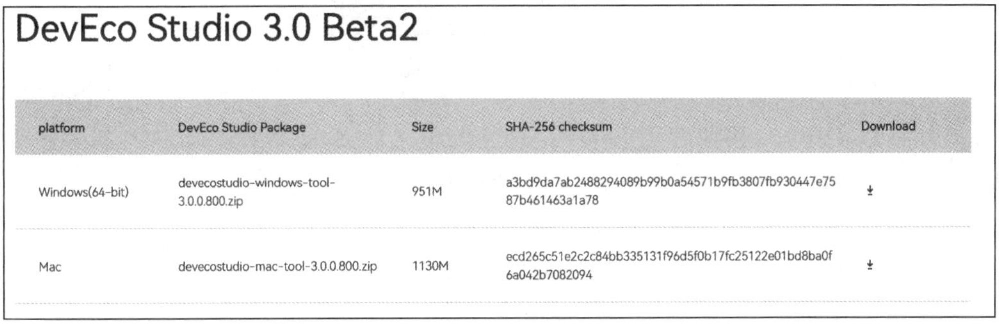

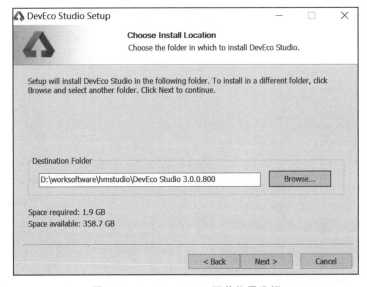

图 1-11　DevEco Studio 下载版本选择

图 1-12　DevEco Studio 下载位置选择

勾选 DevEco Studio 创建桌面快捷方式，之后单击 Next 按钮，如图 1-13 所示。

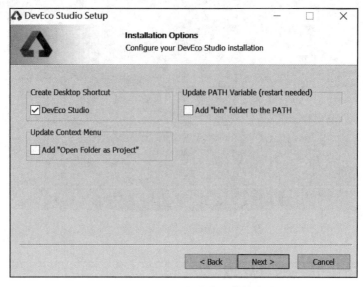

图 1-13　DevEco Studio 创建桌面快捷方式设置

为 DevEco Studio 快捷方式选择一个开始菜单文件夹，也可以手动输入名称来创建新文件夹，填写 Huawei，然后单击 Install 按钮，如图 1-14 所示。

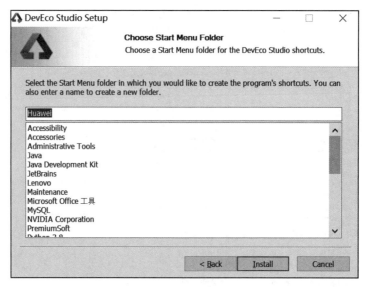

图 1-14　DevEco Studio 菜单文件夹设置

软件开始安装，单击 Show details 按钮可显示安装细节，如图 1-15 所示。

DevEco Studio 安装完毕，勾选 Run DevEco Studio 后，单击 Finish 按钮，即可运行，如图 1-16 所示。

阅读各类条款后，单击 Agree 按钮开始使用 DevEco Studio，如图 1-17 所示。

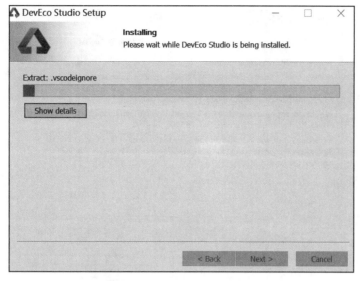

图 1-15　DevEco Studio 正在安装一

图 1-16　DevEco Studio 正在安装二

勾选 Do not import settings，单击 OK 按钮，如图 1-18 所示。

勾选 npm registry，默认填写 npm 配置信息，单击 Start using DevEco Studio 按钮，如图 1-19 所示。

默认勾选 OpenHarmony SDK，选择下载位置，如图 1-20 所示。

检查安装设置，之后单击 Next 按钮，如图 1-21 所示。

选中 Accept，表示同意许可证协议，再单击 Next 按钮，如图 1-22 所示。

这时 HarmonyOS SDK 已经开始安装，等待即可。下载完成后单击 Finish 按钮，如图 1-23 所示。

图 1-17　DevEco Studio 条款同意图

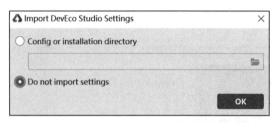

图 1-18　不导入设置

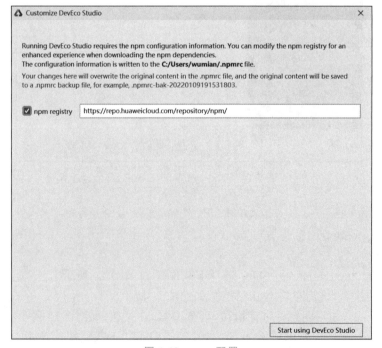

图 1-19　npm 配置

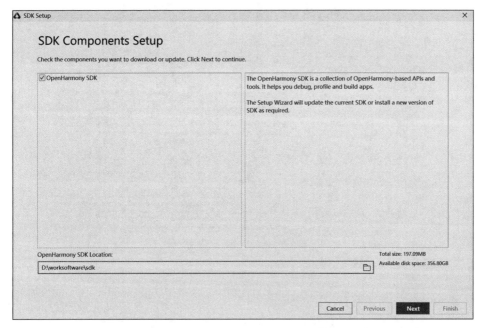

图 1-20　OpenHarmony SDK 下载

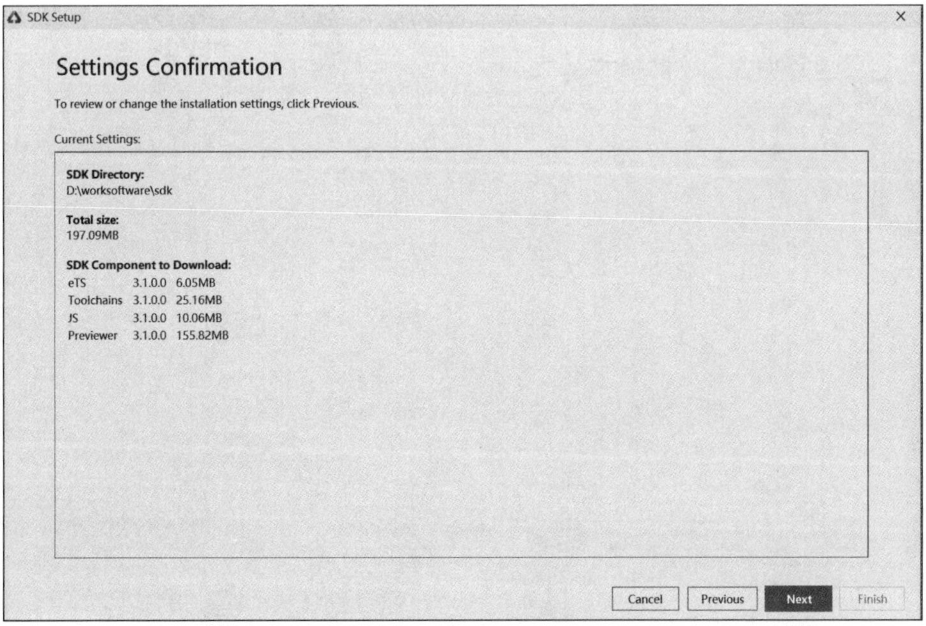

图 1-21　检查安装设置

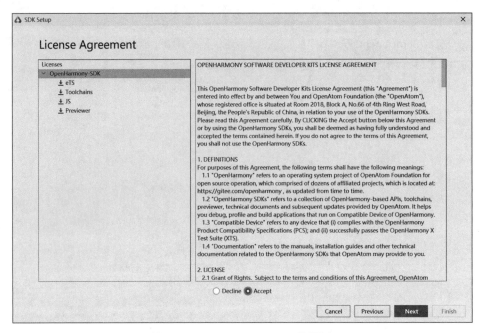

图 1-22　OpenHarmony SDK 许可证

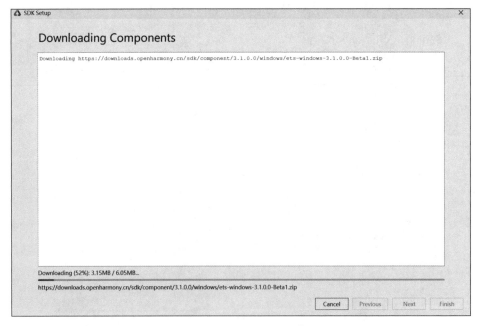

图 1-23　HarmonyOS SDK 下载过程

进入 DevEco Studio 初始界面,选择 Create Project 创建新的项目,如图 1-24 所示。

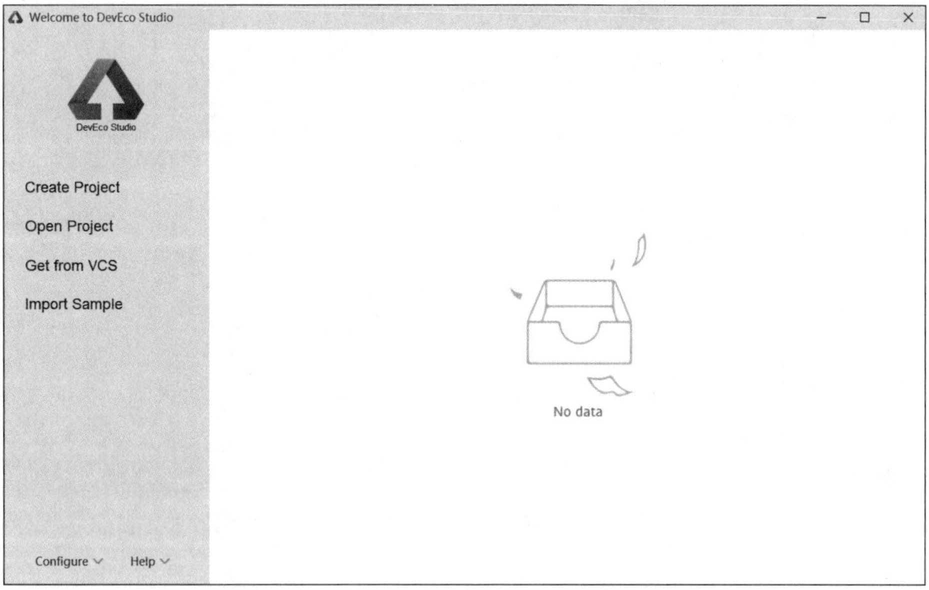

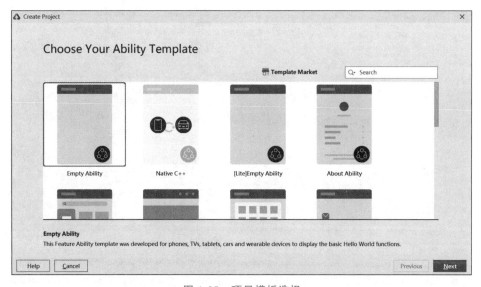

图 1-24　DevEco Studio 初始界面

选择 Empty Ability 创建一个新的空鸿蒙项目模板,再单击 Next 按钮,如图 1-25 所示。

图 1-25　项目模板选择

为了创建一个鸿蒙项目,需要设置 SDK,单击 Install SDK 按钮,如图 1-26 所示。
默认勾选 HarmonyOS Legacy SDK,选择下载位置,如图 1-27 所示。
检查安装设置,单击 Next 按钮进行下一步操作,如图 1-28 所示。

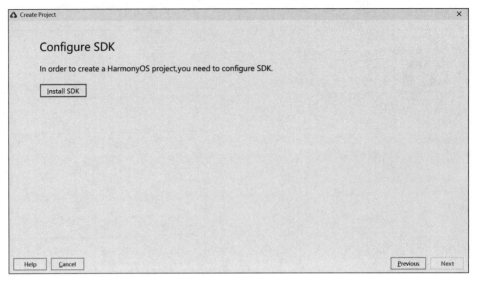

图 1-26　SDK 设置

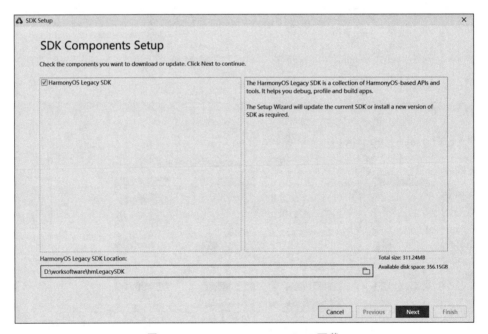

图 1-27　HarmonyOS Legacy SDK 下载

选中 Accept，表示同意许可证协议，再单击 Next 按钮进行下一步操作，如图 1-29 所示。

这时 HarmonyOS SDK 已经开始安装，等待即可，安装完成后单击 Finish 按钮进行下一步操作，如图 1-30 所示。

HarmonyOS SDK 安装成功，单击 Next 按钮进行下一步操作，如图 1-31 所示。

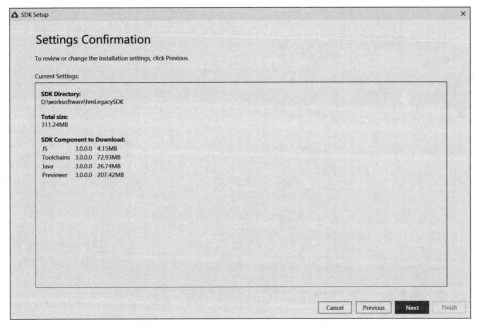

图 1-28　安装设置检查

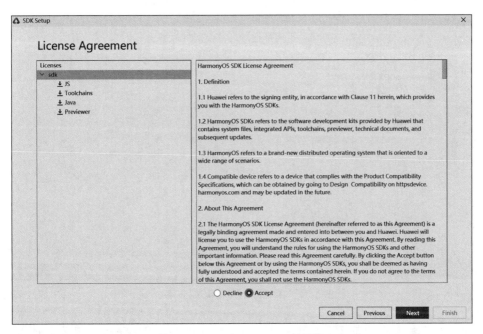

图 1-29　HarmonyOS SDK 许可证

配置项目信息,如项目名、包名、项目类型、保存地址、设备类型等,之后即可进入主界面,如图 1-32 所示。

为了便于查看页面效果,不需要打开远程模拟器,可以使用 Previewer 预览工具,依

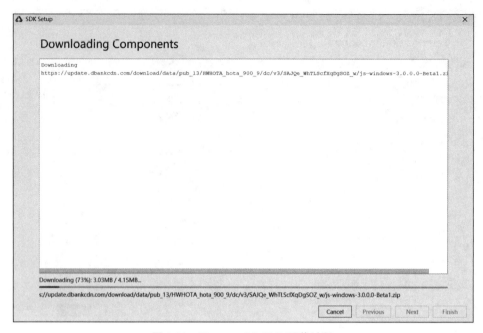

图 1-30　HarmonyOS SDK 下载过程

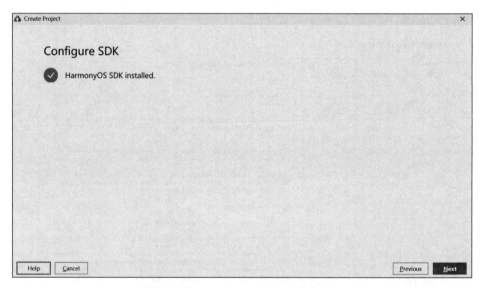

图 1-31　HarmonyOS SDK 安装成功

次打开开发工具左上角的 File＞settings＞DevEco Labs＞Previewer，勾选 Enable Java Previewer 后单击 Apply 再单击 OK 按钮，如图 1-33 所示。

首次开发项目时，会自动下载 Gradle，等待自动下载完成即可，如图 1-34 所示。

依次打开"myapplication1＞entry＞src＞main＞resources＞base＞layout＞ability_main.xml"，再依次打开"View＞Tool Windows＞Previewer"就可以查看 App 页面的显示效果，如图 1-35 所示。

第 1 章　HarmonyOS 简介与环境搭建

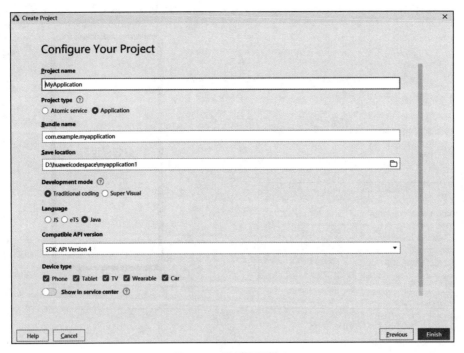

图 1-32　配置项目信息

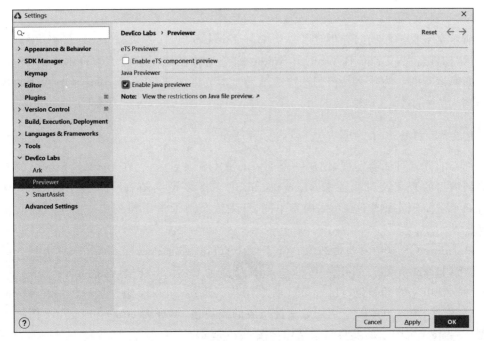

图 1-33　Previewer 设置

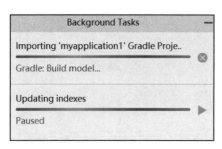

图 1-34　Gradle 下载

图 1-35　使用 Previewer

1.3　HarmonyOS 项目运行过程分析

前面章节只是根据向导创建了一个 HarmonyOS 项目，并未编写任何代码，预览后却能显示"你好,世界"字符串，DevEco Studio 究竟为我们做了些什么？HarmonyOS 程序又是如何运行的？为什么会得到这样的结果？本节将详细介绍 HarmonyOS 程序的执行过程。

1.3.1　HarmonyOS 应用程序结构分析

细心的读者可能会发现，创建一个 HarmonyOS 项目后，会在 DevEco Studio 的左边生成该项目的程序文件。在 DevEco Studio 中支持多种视图查看该程序结构，默认为 Project 视图，不同视图下显示的内容不同，开发人员可通过下拉列表在多个视图间快速切换。

HarmonyOS 应用发布形态为 App Pack（Application Package,App）,它由一个或多个 HAP（HarmonyOS Ability Package）以及描述每个 HAP 属性的 pack.info 组成。HAP 是 Ability 的部署包，HarmonyOS 应用代码围绕 Ability 组件展开。一个 HAP 是由代码、资源、第三方库及应用配置文件组成的模块包，可分为 entry 和 feature 两种模块类型。Project 视图下的项目目录结构如图 1-36 所示。

本节主要介绍一些常用的文件夹，gradle 文件夹中存放着一般情况下不需要进行修改由系统自动生成的 gradle 配置文件。entry 文件夹中存放着应用程序的核心内容，比如代码及各类资源。libs 文件夹中存放着用于存放 entry 模块的依赖文件，如 JAR 包。java 文件夹里存放着用于创建和调整布局以及提供交互功能的 Java 代码。resources 内的 base 文件夹中存放着项目所需的各种资源。element 文件夹中存放着在其他地方可以被引用，可以表示字符串和颜色值等的 JSON 格式的文件。graphic 文件夹中存放着可以自己绘制的 XML 资源文件，例如可以自定义一个特定颜色的矩形按钮。layout 文件夹中存放着页面的布局 XML 资源文件。media 文件夹中存放着各种媒体资源文件，包括图片、音频、视频等资源。test 文件夹中主要存放用于单元测试的功能代码。config.json 文件中存放着项目的配置信息。

1.3.2　HarmonyOS 应用程序运行过程分析

配置好开发环境之后，就可以运行初始项目了。打开 DevEco Studio，单击 Tools 打开设备管理器 Device Manager 启动远程模拟器，如图 1-37 所示。

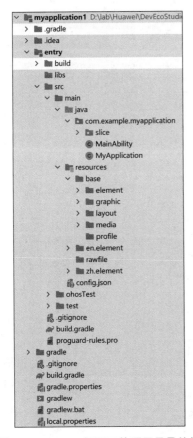

图 1-36　Project 视图下的项目目录结构

图 1-37　Project 视图下的项目目录结构

在左侧我们可以看到有一排远程模拟器可供选择，但是由于我们还没有安装远程模拟器，因此还不能启动远程模拟器设备，单击 Install 按钮，如图 1-38 所示。

这时远程模拟器已经开始安装，等待即可，安装完成后单击 Finish 按钮进行下一步

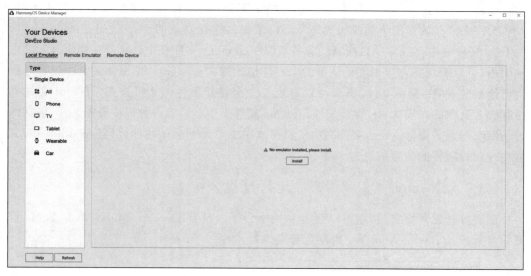

图 1-38　远程模拟器选择面板

操作，如图 1-39 所示。

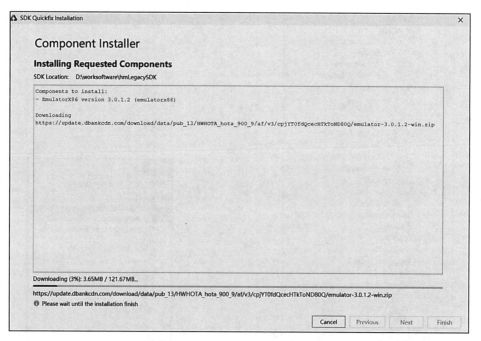

图 1-39　远程模拟器安装

这时的界面没有任何模拟器可选，单击 Remote Emulator 可以看到并没有登录，所以单击 Sign In 按钮进行下一步操作，如图 1-40 所示。

输入之前注册的华为开发者账号、密码，之后单击"登录"按钮，如图 1-41 所示。

登录时需要用户对其进行访问授权，单击"允许"按钮，如图 1-42 所示。

第 1 章　HarmonyOS 简介与环境搭建

图 1-40　远程模拟器选择界面

图 1-41　华为用户登录

图 1-42　华为用户授权

如图1-43所示，登录成功，可以返回 DevEco Studio 进行下一步操作。

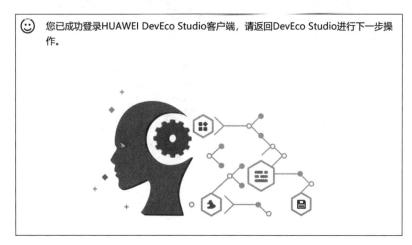

图 1-43　DevEco Studio 登录成功

获取到设备使用权限后，我们可以选择打开设备来启动项目，这里我们发现各个远程模拟器都是就绪状态，选择 Wearable 设备，单击启动，如图 1-44 所示。

Type	Device	Resolution	API	CPU/ABI	Status	Actions
	Mate X2 5G	2480*2200	6	arm	ready	▶
	TV	1920*1080	6	arm	ready	▶
	Wearable	466*466	6	arm	ready	▶
	P40 Pro	1200*2640	7	arm	ready	▶
	P40	1080*2340	6	arm	ready	▶
	MatePad Pro	1600*2560	6	arm	ready	▶

图 1-44　远程模拟器启动

出现初始表盘界面后，单击三角形项目启动 Run 按钮，如图1-45 所示。

图 1-45　项目启动效果

出现"你好,世界"字符串，项目在模拟器上运行成功，如图1-2 所示。

1.4 本章小结

本章介绍了 HarmonyOS 应用开发的基础知识，包括什么是 HarmonyOS、HarmonyOS 的体系结构、HarmonyOS 环境搭建的过程及注意事项、HarmonyOS 程序结构分析、编译的主要流程、运行过程分析。通过本章的学习，读者应重点掌握 HarmonyOS 的环境搭建，包括 DevEco Studio 的安装、HarmonyOS SDK 的下载等，并熟悉 HarmonyOS 应用的创建、运行方式。了解 HarmonyOS 应用程序中各文件夹的作用，掌握 HarmonyOS 编译的主要流程以及掌握 HarmonyOS 程序的运行过程。能够独立搭建 HarmonyOS 开发环境并描述 HarmonyOS 程序的运行过程。

1.5 课后习题

1. HarmonyOS 体系结构大致分为_____、_____、_____和_____四层。
2. HarmonyOS 的底层建立在（　　）操作系统之上。（多选）
   A. Symbian　　　　B. UNIX　　　　C. LiteOS　　　　D. Linux
3. 创建一个 HarmonyOS 项目，该项目的应用名称为 Name，包名为 com.text.book，并实现如图 1-46 所示的运行结果。

图 1-46　界面效果图

第 2 章

HarmonyOS 界面编程基础

课程思政 2

本章要点

- HarmonyOS 界面设计概述
- HarmonyOS 基础界面控件
- HarmonyOS 布局管理器
- HarmonyOS 高级界面控件

本章知识结构图（见图 2-1）

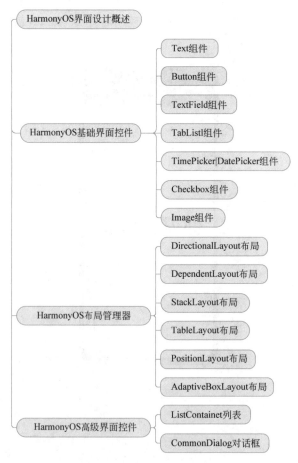

图 2-1　本章知识结构图

本章示例(见图 2-2)

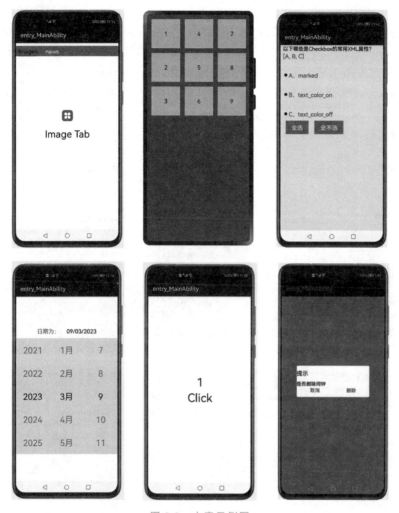

图 2-2　本章示例图

第 1 章通过一个简单的 Hello World 程序介绍了 HarmonyOS 应用程序的运行过程。HarmonyOS 程序开发主要分三部分：界面设计、代码流程控制和数据处理。代码和数据主要由开发者编写和维护，大部分用户是不关心的，展现在用户面前最直观的就是界面设计，作为一个程序设计者，必须首先考虑用户的体验，只有用户满意的开发产品，应用才能推广，才有价值，因此界面设计尤为重要。

HarmonyOS 中提供了丰富的界面控件，开发者熟悉这些控件的功能和用法后，直接调用就可以设计出优秀的图形用户界面。除此之外，HarmonyOS 还允许用户开发自定义的控件，在系统的功能控件基础之上设计出符合自己需求的个性化控件。本章将详细讲解 HarmonyOS 中的一些最基本的控件以及简单的布局管理。通过本章的学习，读者应该能开发出简单的图形用户界面。

2.1 HarmonyOS 界面设计概述

HarmonyOS 界面中所有的组件元素都由 Component 与 ComponentContainer 对象构成，Component 是界面中所有组件的基类，可以将组件分为布局类型、显示类型、交互类型三类，其支持的 XML 属性，其他组件也支持，ComponentContainer 是一个容器（如 DirectionalLayout），没有这个容器，其他组件就不能显示，也不能交互，它不仅可以包含任意个 Component，还可以嵌套其他 ComponentContainer 进行组合。用户可以通过布局容器中的组件对象进行交互操作，如图 2-3 所示。

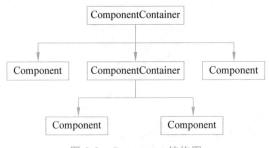

图 2-3 Component 结构图

Component 是所有组件的基类，其常用 XML 属性如表 2-1 所示。

表 2-1 Component XML 属性表

属性名称	说明	使用案例
id	控件 identity，用以识别不同控件对象	ohos:id="＄＋id:text"
width	宽度，必填项	ohos:width="20vp"
hight	高度，必填项	ohos:hight="20vp"
margin	外边距	ohos:margin="20vp"
padding	内边距	ohos:padding="20vp"

2.2 HarmonyOS 基础界面控件

2.2.1 Text 组件

Text 是用来显示一段字符串的常见组件，在初始化界面中的"Hello world"文本就是一个 Text 组件。Text 的共有 XML 属性继承自 Component。Text 组件可以扩展为 Button 组件以及 TextField 组件。

Text 的常用自有 XML 属性如表 2-2 所示。

表 2-2　Text XML 属性表

属性名称	释　　义	使用案例
text	显示文本	ohos:text="你好,世界"
text_color	文本颜色	ohos:text_color="#A8FFFFFF"
italic	文本字体是否为斜体	ohos:italic="true"
hint	提示文本	ohos:hint="联系人"
text_size	文本大小	ohos:text_size="20" ohos:text_size="20fp"
text_font	字体类型	ohos:text_font="serif"
multiple_lines	多行模式设置	ohos:multiple_lines="true"
max_text_lines	文本最大行数	ohos:max_text_lines="2"
text_alignment	文本对齐方式	ohos:text_alignment="top" ohos:text_alignment="horizontal_center|bottom"

打开 src＞main＞resources＞base,可以找到 graphic 文件夹,这是专门用来存放图形的 XML 样式文件。在 graphic 文件夹下新建一个 color_gray.xml,定义一个灰色矩形模板,详细代码如程序清单 2-1 所示。

程序清单 2-1　hmos\ch02_01\Text\entry\src\main\resources\base\garphic\color_gray.xml

```
1   <?xml version="1.0" encoding="utf-8"?>
2   <shape xmlns:ohos="http://schemas.huawei.com/res/ohos"
3          ohos:shape="rectangle">
4     <solid ohos:color="gray"/>
5   </shape>
```

在 layout 文件夹下打开 ability_main.xml,定义一个以灰色矩形为背景的文本,详细代码如程序清单 2-2 所示。

程序清单 2-2　hmos\ch02_01\Text\entry\src\main\resources\base\layout\ability_main.xml

```
1   <?xml version="1.0" encoding="utf-8"?>
2   <DirectionalLayout
3       xmlns:ohos="http://schemas.huawei.com/res/ohos"
4       ohos:height="match_parent"
5       ohos:width="match_parent"
6       ohos:alignment="center"
7       ohos:orientation="vertical">
8       <Text
9           ohos:id="$+id:text"
10          ohos:width="match_content"
11          ohos:height="match_content"
12          ohos:text="Text"
```

```
13          ohos:text_size="80fp"
14          ohos:text_color="black"
15          ohos:left_margin="15vp"
16          ohos:bottom_margin="15vp"
17          ohos:right_padding="15vp"
18          ohos:left_padding="15vp"
19          ohos:background_element="$graphic:color_gray"/>
20      </DirectionalLayout>
```

打开 View>Tool Windows>Previewer 查看页面的显示效果,如图 2-4 所示。

图 2-4　使用 xml 设置背景图

设置字体的大小和颜色,详细代码如程序清单 2-3 所示。

程序清单 2-3：hmos\ch02\01\Text\entry\src\main\resources\base\layout\ability_second.xml

```
1   <?xml version="1.0" encoding="utf-8"?>
2   <DirectionalLayout
3       xmlns:ohos="http://schemas.huawei.com/res/ohos"
4       ohos:height="match_parent"
5       ohos:width="match_parent"
6       ohos:alignment="center"
7       ohos:orientation="vertical">
8       <Text
9           ohos:id="$+id:text"
10          ohos:width="match_content"
11          ohos:height="match_content"
12          ohos:text="Text"
13          ohos:text_size="80fp"
14          ohos:text_color="red"
15          ohos:background_element="$graphic:color_gray"/>
16  </DirectionalLayout>
```

打开 View>Tool Windows>Previewer 查看页面的显示效果,如图 2-5 所示。

图 2-5　字体大小和颜色效果图

设置字体风格和字重的效果,详细代码如程序清单 2-4 所示。

程序清单 2-4：hmos\ch02\01\Text\entry\src\main\resources\base\layout\ability_third.xml

```
1   <?xml version="1.0" encoding="utf-8"?>
2   <DirectionalLayout
3       xmlns:ohos="http://schemas.huawei.com/res/ohos"
4       ohos:height="match_parent"
5       ohos:width="match_parent"
6       ohos:alignment="center"
7       ohos:orientation="vertical">
8       <Text
9           ohos:id="$+id:text"
10          ohos:width="match_content"
11          ohos:height="match_content"
12          ohos:text="Text"
13          ohos:text_size="80fp"
14          ohos:text_color="red"
15          ohos:italic="true"
16          ohos:text_weight="700"
17          ohos:text_font="serif"
18          ohos:left_margin="15vp"
19          ohos:bottom_margin="15vp"
20          ohos:right_padding="15vp"
21          ohos:left_padding="15vp"
22          ohos:background_element="$graphic:color_gray"/>
23  </DirectionalLayout>
```

打开 View＞Tool Windows＞Previewer 查看页面的显示效果，如图 2-6 所示。

图 2-6　字体风格和字重效果图

设置文本的对齐方式，详细代码如程序清单 2-5 所示。

程序清单 2-5：hmos\ch02\01\Text\entry\src\main\resources\base\layout\ability_fourth.xml

```
1   <?xml version="1.0" encoding="utf-8"?>
2   <DirectionalLayout
3       xmlns:ohos="http://schemas.huawei.com/res/ohos"
4       ohos:height="match_parent"
5       ohos:width="match_parent"
6       ohos:alignment="center"
7       ohos:orientation="vertical">
```

```
8       <Text
9           ohos:id="$+id:text"
10          ohos:width="200vp"
11          ohos:height="200vp"
12          ohos:text="Text"
13          ohos:text_size="80fp"
14          ohos:text_color="red"
15          ohos:italic="true"
16          ohos:text_weight="700"
17          ohos:text_font="serif"
18          ohos:text_alignment="horizontal_center|top"
19          ohos:background_element="$graphic:color_gray"/>
20  </DirectionalLayout>
```

打开 View>Tool Windows>Previewer 查看页面的显示效果,如图 2-7 所示。

图 2-7 文本对齐效果图

设置文本换行和最大显示行数,详细代码如程序清单 2-6 所示。

```
1   <?xml version="1.0" encoding="utf-8"?>
2   <DirectionalLayout
3       xmlns:ohos="http://schemas.huawei.com/res/ohos"
4       ohos:height="match_parent"
5       ohos:width="match_parent"
6       ohos:alignment="center"
7       ohos:orientation="vertical">
8       <Text
9           ohos:id="$+id:text"
10          ohos:width="120vp"
11          ohos:height="match_content"
12          ohos:text="TextText"
13          ohos:text_size="40fp"
14          ohos:text_color="red"
15          ohos:italic="true"
16          ohos:text_weight="700"
17          ohos:text_font="serif"
```

```
18          ohos:multiple_lines="true"
19          ohos:max_text_lines="2"
20          ohos:left_margin="15vp"
21          ohos:bottom_margin="15vp"
22          ohos:right_padding="15vp"
23          ohos:left_padding="15vp"
24          ohos:text_alignment="horizontal_center|top"
25          ohos:background_element="$graphic:color_gray"/>
26  </DirectionalLayout>
```

打开 View>Tool Windows>Previewer 查看页面的显示效果,如图 2-8 所示。

2.2.2 Button 组件

Button 是一种单击可以触发对应操作的组件,通常由文本或图标组成,也可以由图标和文本共同组成。Button 无自有的 XML 属性,共有 XML 属性继承自 Text。

图 2-8 文本换行和最大显示效果图

在 layout 目录下的 XML 文件中创建 Button,并设置按钮的背景形状、颜色。常用的背景如文本背景、按钮背景,通常采用 XML 格式放置在 graphic 目录下。

在 Project 窗口,打开 entry>src>main>resources>base,右击 graphic 文件夹,选择 New>File,命名为"background_button.xml",在该文件中定义按钮的背景形状、颜色,详细代码如程序清单 2-7 所示。

程序清单 2-7: hmos\ch02\02\Button\entry\src\main\resources\base\graphic\background_button.xml

```
1  <?xml version="1.0" encoding="utf-8"?>
2  <shape xmlns:ohos="http://schemas.huawei.com/res/ohos"
3      ohos:shape="rectangle">
4  <corners
5      ohos:radius="10"/>
6  <solid
7      ohos:color="blue"/>
8  </shape>
```

在 Project 窗口,打开 entry>src>main>resources>base>media,添加所需图片至 media 目录下。

设置按钮的背景形状及颜色,详细代码如程序清单 2-8 所示。

程序清单 2-8: hmos\ch02\02\Button\entry\src\main\resources\base\layout\ability_main.xml

```
1  <?xml version="1.0" encoding="utf-8"?>
2  <DirectionalLayout
```

```
 3      xmlns:ohos="http://schemas.huawei.com/res/ohos"
 4      ohos:height="match_parent"
 5      ohos:width="match_parent"
 6      ohos:alignment="center"
 7      ohos:orientation="horizontal">
 8      <Image
 9          ohos:height="match_content"
10          ohos:width="match_content"
11          ohos:image_src="$media:button"/>
12      <Button
13          ohos:id="$+id:button"
14          ohos:width="match_content"
15          ohos:height="match_content"
16          ohos:text_size="50fp"
17          ohos:text="单击登录"
18          ohos:background_element="$graphic:background_button"
19          ohos:left_margin="15vp"/>
20      </DirectionalLayout>
```

打开 View>Tool Windows>Previewer 查看具体页面的显示效果,如图 2-9 所示。

图 2-9　按钮的背景形状及颜色效果图

按钮的重要作用是:当用户单击按钮时,Button 对象会收到一个单击事件,程序会执行相应的操作或者界面会出现相应的变化。开发者可以自定义响应单击事件的方法。例如,创建一个 Component.ClickedListener 对象,然后通过调用 setClickedListener 将其分配给按钮。

设置按钮单击事件,当单击 Click 按钮时,上方的数字会自动加一,界面代码片段如程序清单 2-9 所示。

程序清单 2-9: hmos\ch02\02\Button2\entry\src\main\resources\base\layout\ability_main.xml

```
1   <?xml version="1.0" encoding="utf-8"?>
2   <DirectionalLayout
3       xmlns:ohos="http://schemas.huawei.com/res/ohos"
4       ohos:height="match_parent"
5       ohos:width="match_parent"
6       ohos:alignment="center"
7       ohos:orientation="vertical">
8       <Text
9           ohos:id="$+id:text"
10          ohos:height="match_content"
11          ohos:width="match_content"
12          ohos:layout_alignment="horizontal_center"
13          ohos:text="0"
14          ohos:text_size="40vp"/>
15      <Button
16          ohos:id="$+id:button"
17          ohos:height="match_content"
18          ohos:width="match_content"
19          ohos:layout_alignment="horizontal_center"
20          ohos:text="Click"
21          ohos:text_size="40vp"/>
22  </DirectionalLayout>
```

功能代码片段如程序清单 2-10 所示。

程序清单 2-10: hmos\ch02\02\Button2\entry\src\main\java\com\example\button02\slice\MainAbilitySlice.java

```
1   public class MainAbilitySlice extends AbilitySlice {
2       int i = 0;
3       @Override
4       public void onStart(Intent intent) {
5           super.onStart(intent);
6           super.setUIContent(ResourceTable.Layout_ability_main);
7           Button button = (Button) findComponentById(ResourceTable.Id_button);
8           Text text = (Text) findComponentById(ResourceTable.Id_text);
9           if (button != null) {
10              button.setClickedListener(component -> {
```

```
11              i++;
12              text.setText(""+i);
13          });
14      }
15  }
16 }
```

打开 Tools>Device Manager，登录账号后运行 phone 远程模拟器查看具体页面的显示效果，如图 2-10 所示。

(a) 单击前效果　　　　　(b) 单击后效果

图 2-10　单击 Click 按钮效果

2.2.3　TextField 组件

TextField 提供了一种文本输入框用来接收用户输入的内容。TextField 的共有 XML 属性继承自 Text 组件，常用的自有 XML 属性如表 2-3 所示。

表 2-3　TextField XML 属性表

属性名称	说　　明	使 用 案 例
basement	输入框基线	ohos：basement="#000000"

为展示 TextField 的使用，在 layout 目录下的 ability_main.xml 文件中创建一个 TextField 组件，详细代码片段如程序清单 2-11 所示。

程序清单 2-11：HmosCh02_03_TextField\entry\src\main\resources\base\layout\ability_main.xml

```
1   <?xml version="1.0" encoding="utf-8"?>
```

```
 2  <DirectionalLayout
 3    xmlns:ohos="http://schemas.huawei.com/res/ohos"
 4      ohos:height="match_parent"
 5      ohos:width="match_parent"
 6      ohos:alignment="center"
 7      ohos:orientation="vertical">
 8      <TextField
 9          ohos:id="$+id:text_field"
10          ohos:height="match_content"
11          ohos:width="match_content"
12          ohos:text_size="30fp"
13          ohos:hint="输入账号或邮箱"
14          ohos:text_alignment="vertical_center"
15          ohos:basement="black"/>
16  </DirectionalLayout>
```

打开 View>Tool Windows>Previewer 查看具体页面的显示效果,如图 2-11 所示。

图 2-11　TextField 输入框基线

2.2.4　TabList 组件

TabList 的共有 XML 属性继承自 ScrollView。TabList 常用的自有 XML 属性如表 2-4 所示。

表 2-4　TabList 常用的自有 XML 属性

属性名称	说　　明	使 用 案 例
basement	输入框基线	ohos:basement="♯000000"
orientation	页签排列方向	ohos:orientation="horizontal" ohos:orientation="vertical"
normal_text_color	未选中的文本颜色	ohos:normal_text_color="♯FFFFFFFF"
selected_text_color	选中的文本颜色	ohos:selected_text_color="♯FFFFFFFF"
selected_tab_indicator_color	选中页签的颜色	ohos:selected_tab_indicator_color="♯FFFFFFFF"
selected_tab_indicator_height	选中页签的高度	ohos:selected_tab_indicator_height="20vp"

当进入 App 首页的时候会展示给用户一个框架，比如微信，展示了四个 Tab，分别对应不同的板块（微信、通讯录、发现、我），不同的板块可以相互切换。接下来学习如何在 HarmonyOS 中实现 TabList。TabList 可以实现多个页签栏的切换，Tab 为某个页签。子页签通常放在内容区上方，展示不同的分类。页签名称应该简洁、明了，清晰描述分类的内容。原理：页面上方设置一个 TabList，下面设置一个 ComponentContainer 作为容器。TabList 上设置几个 Tab，单击 Tab 的时候就将不同的 Tab Layout 载入 ComponentContainer 中展现，如图 2-12 所示。

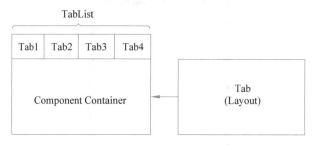

图 2-12　TabList 组成构造图

接下来将用代码实现如下界面，该界面包含 Images 和 news 两个 Tab，可以相互切换，如图 2-13 所示。

TabList 项目文件结构图如图 2-14 所示。

图 2-13　Image Tab 界面

图 2-14　TabList 项目文件结构图

首先准备 TabList 页面布局,在 layout 目录下的 ability_main.xml 文件中创建一个 TabList 组件,详细代码片段如程序清单 2-12 所示。

程序清单 2-12: hmos\ch02\04\TabList\entry\src\main\resources\base\layout\ability_main.xml

```xml
1   <?xml version="1.0" encoding="utf-8"?>
2   <DirectionalLayout
3       xmlns:ohos="http://schemas.huawei.com/res/ohos"
4       ohos:height="match_parent"
5       ohos:width="match_parent"
6       ohos:orientation="vertical">
7       <TabList
8           ohos:id="$+id:tab_list"
9           ohos:background_element="gray"
10          ohos:top_margin="10vp"
11          ohos:tab_margin="24vp"
12          ohos:tab_length="140vp"
13          ohos:text_size="20fp"
14          ohos:height="36vp"
15          ohos:width="match_parent"
16          ohos:layout_alignment="center"
17          ohos:orientation="horizontal"
18          ohos:text_alignment="center"
19          ohos:normal_text_color="white"
20          ohos:selected_text_color="black"
21          ohos:selected_tab_indicator_color="black"
22          ohos:selected_tab_indicator_height="2vp"/>
23      <DirectionalLayout
24          ohos:id="$+id:tab_container"
25          ohos:height="match_parent"
26          ohos:width="match_parent">
27      </DirectionalLayout>
28  </DirectionalLayout>
```

ability_main.xml 布局代码中的第 7~22 行用于生成 TabList 组件,定义了 TabList 的基本属性;第 23~27 行用于生成 Tab 页面的容器,然后准备 Images 页面。Images 页面主要包含一个 Image 文件(开发工具图标)和简单的文字(Image Tab)表示,详细代码片段如程序清单 2-13 所示。

程序清单 2-13: hmos\ch02\04\TabList\entry\src\main\resources\base\layout\tabcontent_Image.xml

```xml
1   <?xml version="1.0" encoding="utf-8"?>
2   <DirectionalLayout
3       xmlns:ohos="http://schemas.huawei.com/res/ohos"
```

```
4        ohos:height="match_parent"
5        ohos:width="match_parent"
6        ohos:orientation="vertical">
7        <Component
8            ohos:height="0vp"
9            ohos:weight="3"
10           ohos:width="match_parent"/>
11       <DirectionalLayout
12           xmlns:ohos="http://schemas.huawei.com/res/ohos"
13           ohos:height="match_content"
14           ohos:width="match_content"
15           ohos:layout_alignment="center"
16           ohos:orientation="vertical">
17           <Image
18               ohos:id="$+id:image"
19               ohos:width="match_content"
20               ohos:height="match_content"
21               ohos:layout_alignment="center"
22               ohos:image_src="$media:icon"/>
23           <Component
24               ohos:height="20vp"
25               ohos:width="match_parent"/>
26           <Text
27               ohos:id="$+id:text_helloworld"
28               ohos:height="match_content"
29               ohos:width="match_content"
30               ohos:layout_alignment="horizontal_center"
31               ohos:text="Image Tab"
32               ohos:text_color="black"
33               ohos:text_size="100"/>
34       </DirectionalLayout>
35       <Component
36           ohos:height="0vp"
37           ohos:weight="5"
38           ohos:width="match_parent"/>
39   </DirectionalLayout>
```

接着准备 News 界面，该界面与 Images 界面相似，具体代码在 tabcontent_New.xml 中实现，详细代码片段如程序清单 2-14 所示。

程序清单 2-14：hmos\ch02\04\TabList\entry\src\main\resources\base\layout\tabcontent_New.xml

```
1    <?xml version="1.0" encoding="utf-8"?>
2    <DirectionalLayout
```

```
3      xmlns:ohos="http://schemas.huawei.com/res/ohos"
4      ohos:height="match_parent"
5      ohos:width="match_parent"
6      ohos:orientation="vertical">
7      <Component
8          ohos:height="0vp"
9          ohos:weight="3"
10         ohos:width="match_parent"/>
11     <DirectionalLayout
12         xmlns:ohos="http://schemas.huawei.com/res/ohos"
13         ohos:height="match_content"
14         ohos:width="match_content"
15         ohos:layout_alignment="center"
16         ohos:orientation="vertical">
17     <Component
18         ohos:height="20vp"
19         ohos:width="match_parent"/>
20     <Text
21         ohos:id="$+id:text_helloworld"
22         ohos:height="match_content"
23         ohos:width="match_content"
24         ohos:layout_alignment="horizontal_center"
25         ohos:text="New Tab"
26         ohos:text_color="black"
27         ohos:text_size="100"/>
28     </DirectionalLayout>
29     <Component
30         ohos:height="0vp"
31         ohos:weight="5"
32         ohos:width="match_parent"/>
33 </DirectionalLayout>
```

最后整合两个 Tab 生成 TabList 画面,详细代码片段如程序清单 2-15 所示。

程序清单 2-15: hmos\ch02\04\TabList\entry\src\main\java\com\example\tablist\slice\MainAbilitySlice.java

```
1  public class MainAbilitySlice extends AbilitySlice {
2      private Component imageContent;
3      private Component newContent;
4      private FrameAnimationElement frameAnimationElement;
5      @Override
6      public void onStart(Intent intent) {
7          super.onStart(intent);
```

```
8         super.setUIContent(ResourceTable.Layout_ability_main);
9         TabList tabList = (TabList) findComponentById(ResourceTable.Id_tab_list);
10        tabList.setTabLength(200);            //设置 Tab 的宽度
11        tabList.setTabMargin(26);             //设置两个 Tab 之间的间距
12        TabList.Tab tab1 = tabList.new Tab(getContext());
13        tab1.setText("Images");
14        tabList.addTab(tab1);
15        TabList.Tab tab2 = tabList.new Tab(getContext());
16        tab2.setText("news");
17        tabList.addTab(tab2);
18        AbilitySlice slice = this;
19        tabList.addTabSelectedListener(new TabList.TabSelectedListener() {
20            @Override
21            public void onSelected(TabList.Tab tab) {
22            ComponentContainer container = (ComponentContainer) findComponentById(ResourceTable.Id_tab_container);
23                if(tab.getText().equals("Images")) {
24                    imageContent = LayoutScatter.getInstance(slice).parse(ResourceTable.Layout_tabcontent_Image, null, false);
25                    container.addComponent(imageContent);
26                }
27            else
28                {
29                    newContent = LayoutScatter.getInstance(slice).parse(ResourceTable.Layout_tabcontent_New, null, false);
30                    container.addComponent(newContent);
31                }
32            }
33            @Override
34            public void onUnselected(TabList.Tab tab) {
35            ComponentContainer container = (ComponentContainer) findComponentById(ResourceTable.Id_tab_container);
36                container.removeAllComponents();
37            }
38            @Override
39            public void onReselected(TabList.Tab tab) {
40            ComponentContainer container = (ComponentContainer) findComponentById(ResourceTable.Id_tab_container);
41                if(tab.getText().equals("Images")) {
42                    container.addComponent(imageContent);
43                }
44            else
```

```
45                    {
46                        container.addComponent(newContent);
47                    }
48                }
49            });
50    //最开始选择 tab1
51            tabList.selectTab(tab1);
52        }
53    }
```

单击不同的 Tab,界面会进行切换,单击 news 时会显示 New Tab,单击 Images 时会显示 Image Tab,如图 2-15 所示。

图 2-15 New Tab 界面

2.2.5 DatePicker 组件

DatePicker 主要供用户选择日期。DatePicker 的共有 XML 属性继承自 StackLayout (StackLayout 在本章 2.3.3 节有详细介绍)。DatePicker 的常用自有 XML 属性如表 2-5 所示。

表 2-5 DatePicker 的常用自有 XML 属性

属性名称	说明	使用案例
basement	输入框基线	ohos:basement="#000000"
date_order	显示格式,年月日	ohos:date_order="day-month-year"

续表

属性名称	说 明	使用案例
day_fixed	日期是否固定	ohos:day_fixed="true"
month_fixed	月份是否固定	ohos:month_fixed="true"
year_fixed	年份是否固定	ohos:year_fixed="true"
max_date	最大日期	ohos:max_date="1627747200"
min_date	最小日期	ohos:min_date="1627747200"
normal_text_size	未选中文本的大小	ohos:normal_text_size="30"
selected_text_size	选中文本的大小	ohos:selected_text_size="30"
normal_text_color	未选中文本的颜色	ohos:normal_text_color="black"
selected_text_color	选中文本的颜色	ohos:selected_text_color="black"
selector_item_num	显示的项目数量	ohos:selector_item_num="10"
shader_color	着色器的颜色	ohos:shader_color="blue"
wheel_mode_enabled	选择轮是否循环显示数据	ohos:wheel_mode_enabled="true"

使用 DatePicker 要在 XML 中创建 DatePicker，具体代码在 ability_main.xml 中实现，代码片段如程序清单 2-16 所示。

程序清单 2-16：hmos\ch02\05\DatePicker\entry\src\main\resources\base\layout\ability_main.xml

```
1   <?xml version="1.0" encoding="utf-8"?>
2   <DirectionalLayout
3       xmlns:ohos="http://schemas.huawei.com/res/ohos"
4       ohos:height="match_parent"
5       ohos:width="match_parent"
6       ohos:alignment="center"
7       ohos:orientation="vertical">
8       <DatePicker
9           ohos:id="$+id:date_pick"
10          ohos:height="400fp"
11          ohos:width="match_parent"
12          ohos:text_size="30fp"
13          ohos:background_element="#E6E6FA">
14      </DatePicker>
15  </DirectionalLayout>
```

打开 View>Tool Windows>Previewer 查看具体页面的显示效果，如图 2-16 所示。

使用代码获取当前选择的日期，日/月/年，DatePicker 默认选择当前日期。同时使用 HiLog 日志打印出当前日期，功能代码片段如程序清单 2-17 所示。

第 2 章 HarmonyOS 界面编程基础

图 2-16　DatePicker 效果图

```
1   static final HiLogLabel LABEL = new HiLogLabel(HiLog.LOG_APP, 0x00201, "MY_TAG");
2   @Override
3   public void onStart(Intent intent) {
4       super.onStart(intent);
5       super.setUIContent(ResourceTable.Layout_ability_main);
6   //获取 DatePicker 实例
7       DatePicker datePicker = (DatePicker)findComponentById
    (ResourceTable.Id_date_pick);
8       int day = datePicker.getDayOfMonth();
9       int month = datePicker.getMonth();
10      int year = datePicker.getYear();
11      HiLog.info(LABEL, "----"+day+"---"+month+"----"+year);
12  }
```

使用日志(日志的使用方法在本书 6.4 节有详细介绍)打印通过 DatePicker 选择的日期如图 2-17 所示。

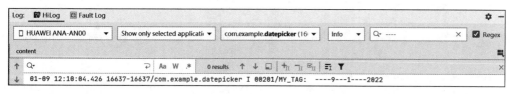

图 2-17　使用日志打印通过 DatePicker 选择的日期

响应日期改变事件：在 ability_main.xml 的 DirectionalLayout 中添加一个 DirectionalLayout，在里面再添加 Text 显示选择日期，代码片段如程序清单 2-18 所示。

程序清单 2-18：hmos\ch02\05\DatePicker2\entry\src\main\resources\base\layout\ability_main.xml

```
1   <DirectionalLayout
2       ohos:height="match_content"
3       ohos:width="match_content"
4       ohos:orientation="horizontal">
5       <Text
6           ohos:id="$+id:guide_date"
7           ohos:height="match_content"
8           ohos:width="match_parent"
9           ohos:layout_alignment="left"
10          ohos:hint="日期为:"
11          ohos:margin="8vp"
12          ohos:padding="4vp"
13          ohos:text_size="14fp"/>
14      <Text
15          ohos:id="$+id:text_date"
16          ohos:height="match_content"
17          ohos:width="match_parent"
18          ohos:layout_alignment="right"
19          ohos:hint="date"
20          ohos:margin="8vp"
21          ohos:padding="4vp"
22          ohos:text_size="14fp"/>
23  </DirectionalLayout>
```

在 Java 代码中响应日期改变事件，代码片段如程序清单 2-19 所示。

程序清单 2-19：hmos\ch02\05\DatePicker2\entry\src\main\java\com\example\datepicker2\slice\MainAbilitySlice.java

```
1   public void onStart(Intent intent) {
2       super.onStart(intent);
3       super.setUIContent(ResourceTable.Layout_ability_main);
4       //获取 DatePicker 实例
5       DatePicker datePicker = (DatePicker) findComponentById(ResourceTable.Id_date_pick);
6       int day = datePicker.getDayOfMonth();
7       int month = datePicker.getMonth();
8       int year = datePicker.getYear();
9       Text selectedDate = (Text) findComponentById(ResourceTable.Id_text_date);
```

```
10      datePicker.setValueChangedListener(
11          new DatePicker.ValueChangedListener() {
12              @Override
13              public void onValueChanged(DatePicker datePicker, int year,
    int monthOfYear, int dayOfMonth) {
14                  selectedDate.setText(String.format("%02d/%02d/%4d",
    dayOfMonth, monthOfYear, year));
15              }
16          }
17      );
18  }
```

DatePicker 时间改变效果如图 2-18 所示。

(a) 时间改变前　　　　　(b) 时间改变后

图 2-18　DatePicker 时间改变效果图

2.2.6　TimePicker 组件

TimePicker 是供用户选择时间的组件。TimePicker 的共有 XML 属性继承自 StackLayout。TimePicker 的常用自有 XML 属性如表 2-6 所示。

表 2-6　TimePicker 的常用自有 XML 属性

属性名称	说　明	使用案例
mode_24_hour	是否 24 小时制显示	ohos:mode_24_hour="true"
hour	显示小时	ohos:hour="8"
minute	显示分钟	ohos:minute="59"

续表

属性名称	说　　明	使 用 案 例
second	显示秒	ohos:second="59"
text_am	上午文本	ohos:text_am="8:00:00"
text_pm	下午文本	ohos:text_pm="22:00:00"

此外，normal_text_size(未选中文本的大小)、selected_text_size(选中文本的大小)、normal_text_color(未选中文本的颜色)、selected_text_color(选中文本的颜色)、shader_color(着色器的颜色)、selector_item_num(显示的项目数量)、selected_normal_text_margin_ratio(已选文本边距与常规文本边距的比例)、top_line_element(选中项的顶行)、bottom_line_element(选中项的底线)、wheel_mode_enabled(选择轮是否循环显示数据)的使用方法与 DatePicker 相同。

设置时间，界面代码如程序清单 2-20 所示。

程序清单 2-20 huno\ch02\05\TimePicker\entry\src\main\resources\base\layout\ability_main.xml

```
1   <?xml version="1.0" encoding="utf-8"?>
2   <DirectionalLayout
3       xmlns:ohos="http://schemas.huawei.com/res/ohos"
4       ohos:height="match_parent"
5       ohos:width="match_parent"
6       ohos:alignment="center"
7       ohos:orientation="vertical">
8       <TimePicker
9           ohos:id="$+id:time_picker"
10          ohos:height="match_content"
11          ohos:width="match_parent" />
12  </DirectionalLayout>
```

设置时间，代码如程序清单 2-21 所示。

程序清单 2-21 huno\ch02\05\TimePicker\entry\src\main\java\com\example\timepicker\slice\MainAbilitySlice.java

```
1   public void onStart(Intent intent) {
2       super.onStart(intent);
3       super.setUIContent(ResourceTable.Layout_ability_main);
4       TimePicker timePicker = (TimePicker) findComponentById(ResourceTable.Id_time_picker);
5       int hour = timePicker.getHour();
6       int minute = timePicker.getMinute();
7       int second = timePicker.getSecond();
8       timePicker.setHour(8);
```

```
9    timePicker.setMinute(8);
10   timePicker.setSecond(8);
11 }
```

设置时间效果如图 2-19 所示。

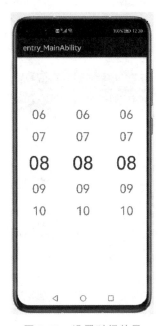

图 2-19　设置时间效果

响应时间改变事件,当小时数大于 8 时,HiLog 打印日志显示"Get up",代码如程序清单 2-22 所示。

程序清单 2-22　hiios\ch02\05\TimePicker2\entry\src\main\java\com\example\timepicker2\slice\MainAbilitySlice.java

```
1    static final HiLogLabel LABEL = new HiLogLabel(HiLog.LOG_APP, 0x00201, "MY_TAG");
2    @Override
3    public void onStart(Intent intent) {
4        super.onStart(intent);
5        super.setUIContent(ResourceTable.Layout_ability_main);
6        TimePicker timePicker = (TimePicker) findComponentById(ResourceTable.
  Id_time_picker);
7        int hour = timePicker.getHour();
8        int minute = timePicker.getMinute();
9        int second = timePicker.getSecond();
10       timePicker.setTimeChangedListener(new TimePicker.TimeChangedListener() {
11           @Override
```

```
12    public void onTimeChanged(TimePicker timePicker, int hour, int minute,
   int second) {
13        if (hour > 8) {
14            HiLog.info(LABEL, "Get up");
15        }
16    }
```

把时间手动设置为大于 8，如图 2-20 所示。

图 2-20　TimePicker 手动设置

在远程模拟器上运行项目，在 HiLog 选项卡中查看日志输出情况（可以在放大镜边的输入框内输入"Get up"快速获取相关信息），如图 2-21 所示。

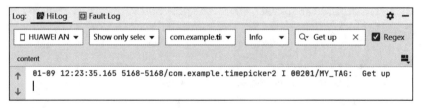

图 2-21　HiLog 打印日志

2.2.7　Checkbox 组件

Checkbox 可以实现选中和取消选中的功能。Checkbox 的共有 XML 属性继承自 Text。Checkbox 的常用自有 XML 属性如表 2-7 所示。

表 2-7 Checkbox 的常用自有 XML 属性

属性名称	说 明	使用案例
marked	当前状态（选中或取消选中）	ohos:marked="true"
text_color_on	处于选中状态的文本颜色	ohos:text_color_on="black"
text_color_off	处于未选中状态的文本颜色	ohos:text_color_off="black"
check_element	状态标志样式	ohos:check_element="$color:black"

设置 Checkbox 在 XML 中配置 Checkbox 的选中和取消选中的状态标志样式。layout 目录下的 background_ability_main.xml 代码如程序清单 2-23 所示。

程序清单 2-23：hmos\ch02\06\Checkbox\entry\src\main\resources\base\layout\background_ability_main.xml

```
1   <?xml version="1.0" encoding="utf-8"?>
2   <DirectionalLayout
3       xmlns:ohos="http://schemas.huawei.com/res/ohos"
4       ohos:height="match_parent"
5       ohos:width="match_parent"
6       ohos:padding="80vp"
7       ohos:background_element="#FFF0AC"
8       ohos:orientation="vertical">
9       <Checkbox
10          ohos:id="$+id:check_box1"
11          ohos:height="match_content"
12          ohos:width="match_content"
13          ohos:text="这是A复选框"
14          ohos:text_size="25fp" />
15      <Checkbox
16          ohos:id="$+id:check_box2"
17          ohos:height="match_content"
18          ohos:width="match_content"
19          ohos:text="这是B复选框"
20          ohos:top_margin="30vp"
21          ohos:check_element="$graphic:checkbox_check_element"
22          ohos:text_size="25fp" />
23  </DirectionalLayout>
```

在 graphic 文件夹下设置 3 个 xml 代码，如图 2-22 所示。
checkbox_check_element.xml 代码，如程序清单 2-24 所示。

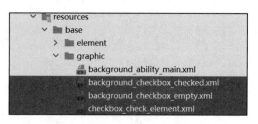

图 2-22 graphic 文件结构

```
1    <? xml version="1.0" encoding="utf-8"?>
2    <state-container xmlns:ohos="http://schemas.huawei.com/res/ohos">
3     < item ohos: state = "component _ state _ checked" ohos: element = " $graphic:
      background_checkbox_checked"/>
4     < item ohos: state = "component _ state _ empty" ohos: element = " $graphic:
      background_checkbox_empty"/>
5    </state-container>
```

background_checkbox_empty.xml 代码,如程序清单 2-25 所示。

```
1    <? xml version="1.0" encoding="UTF-8" ?>
2    <shape xmlns:ohos="http://schemas.huawei.com/res/ohos"
3      ohos:shape="rectangle">
4    <solid
5      ohos:color="#FFFFFF"/>
6    </shape>
```

background_checkbox_checked.xml 代码,如程序清单 2-26 所示。

```
1    <? xml version="1.0" encoding="UTF-8" ?>
2    <shape xmlns:ohos="http://schemas.huawei.com/res/ohos"
3      ohos:shape="rectangle">
4    <solid
5      ohos:color="#FF0000"/>
6    </shape>
```

打开 Tools＞Device Manager,登录账号后,运行远程模拟器查看具体页面的显示效果,如图 2-23 所示。

全选、全不选案例实现:首先在 layout 目录下新建一个 XML 布局文件 demo2_checkbox.xml,代码如程序清单 2-27 所示。

图 2-23　复选框效果

程序清单 2-27　\entry\src\main\resources\base\layout\demo2_checkbox.xml

```xml
1   <?xml version="1.0" encoding="utf-8"?>
2   <DirectionalLayout
3       xmlns:ohos="http://schemas.huawei.com/res/ohos"
4       ohos:height="match_parent"
5       ohos:width="match_parent"
6       ohos:orientation="vertical"
7       ohos:background_element="#eeeeee"
8       ohos:left_padding="10vp"
9       ohos:right_padding="10vp">
10      <Text
11          ohos:height="match_content"
12          ohos:width="match_content"
13          ohos:text_size="18fp"
14          ohos:text="以下哪些是 Checkbox 的常用 XML 属性?"/>
15      <Text
16          ohos:id="$+id:text_answer"
17          ohos:height="match_content"
```

```
18          ohos:width="match_content"
19          ohos:text_size="20fp"
20          ohos:text_color="#FF0000"
21          ohos:text="[]" />
22      <Checkbox
23          ohos:id="$+id:check_box_1"
24          ohos:top_margin="40vp"
25          ohos:height="match_content"
26          ohos:width="match_content"
27          ohos:text="A、marked"
28          ohos:text_size="20fp"
29          ohos:text_color_on="#FF0000"
30          ohos:text_color_off="#000000"/>
31      <Checkbox
32          ohos:id="$+id:check_box_2"
33          ohos:top_margin="40vp"
34          ohos:height="match_content"
35          ohos:width="match_content"
36          ohos:text="B、text_color_on"
37          ohos:text_size="20fp"
38          ohos:text_color_on="#FF0000"
39          ohos:text_color_off="#000000"/>
40      <Checkbox
41          ohos:id="$+id:check_box_3"
42          ohos:top_margin="40vp"
43          ohos:height="match_content"
44          ohos:width="match_content"
45          ohos:text="C、text_color_off"
46          ohos:text_size="20fp"
47          ohos:text_color_on="#FF0000"
48          ohos:text_color_off="#000000" />
49      <DirectionalLayout
50          ohos:height="match_content"
51          ohos:width="match_parent"
52          ohos:orientation="horizontal">
53          <Button
54              ohos:id="$+id:btn1"
55              ohos:height="match_content"
56              ohos:width="match_content"
57              ohos:text="全选"
58              ohos:text_size="20fp"
59              ohos:left_padding="20vp"
60              ohos:right_padding="20vp"
```

```
61              ohos:top_padding="10vp"
62              ohos:bottom_padding="10vp"
63              ohos:margin="10vp"
64              ohos:background_element="blue"/>
65         <Button
66              ohos:id="$+id:btn2"
67              ohos:height="match_content"
68              ohos:width="match_content"
69              ohos:text="全不选"
70              ohos:text_size="20fp"
71              ohos:left_padding="20vp"
72              ohos:right_padding="20vp"
73              ohos:top_padding="10vp"
74              ohos:bottom_padding="10vp"
75              ohos:margin="10vp"
76              ohos:background_element="blue"/>
77     </DirectionalLayout>
78 </DirectionalLayout>
```

在 slice 目录下新建一个 Slice 文件 SecondAbilitySlice.java，为了直接看到该功能页面将 MainAbility.java 中的 super.setMainRoute() 方法参数修改为 SecondAbilitySlice.class.getName()，使用 HashSet 哈希表中的 add() 方法添加数据，使用 remove() 方法删除数据，从而实现选项数据的管理，同时要注意 HashSet 是一个不允许有重复元素的集合。代码如程序清单 2-28 所示。

程序清单 2-28：hmos\ch02\06\Checkbox1\entry\src\main\java\com\example\checkbox\slice\MainAbilitySlice.java

```
1  public class SecondAbilitySlice extends AbilitySlice{
2      private Set<String> selectedSet = new HashSet<>();
3      @Override
4      protected void onStart(Intent intent) {
5          super.onStart(intent);
6          super.setUIContent(ResourceTable.Layout_demo2_checkbox);
7   Checkbox checkbox1 = (Checkbox) findComponentById(ResourceTable.Id_check_box_1);
8      Checkbox checkbox2 = (Checkbox) findComponentById(ResourceTable.Id_check_box_2);
9      Checkbox checkbox3 = (Checkbox) findComponentById(ResourceTable.Id_check_box_3);
10 checkbox1.setCheckedStateChangedListener(new AbsButton.CheckedStateChangedListener() {
11         @Override
```

```
12          public void onCheckedChanged(AbsButton absButton, boolean b) {
13              if (b) {
14                  selectedSet.add("A");
15              } else {
16                  selectedSet.remove("A");
17              }
18              showAnswer();
19          }
20      });
21  checkbox2.setCheckedStateChangedListener(new AbsButton.
    CheckedStateChangedListener() {
22          @Override
23          public void onCheckedChanged(AbsButton absButton, boolean b) {
24              if (b) {
25                  selectedSet.add("B");
26              } else {
27                  selectedSet.remove("B");
28              }
29              showAnswer();
30          }
31  });
32  checkbox3.setCheckedStateChangedListener(new AbsButton.
    CheckedStateChangedListener() {
33          @Override
34          public void onCheckedChanged(AbsButton absButton, boolean b) {
35              if (b) {
36                  selectedSet.add("C");
37              } else {
38                  selectedSet.remove("C");
39              }
40              showAnswer();
41          }
42  });
43      Button btn1 = (Button) findComponentById(ResourceTable.Id_btn1);
44      Button btn2 = (Button) findComponentById(ResourceTable.Id_btn2);
45      btn1.setClickedListener(new Component.ClickedListener() {
46          @Override
47          public void onClick(Component component) {
48              checkbox1.setChecked(true);
49              checkbox2.setChecked(true);
50              checkbox3.setChecked(true);
51              selectedSet.add("A");
52              selectedSet.add("B");
53              selectedSet.add("C");
54              showAnswer();
```

```
55                    }
56                });
57         btn2.setClickedListener(new Component.ClickedListener() {
58             @Override
59             public void onClick(Component component) {
60                 checkbox1.setChecked(false);
61                 checkbox2.setChecked(false);
62                 checkbox3.setChecked(false);
63                 selectedSet.remove("A");
64                 selectedSet.remove("B");
65                 selectedSet.remove("C");
66                 showAnswer();
67             }
68         });
69     }
70     private void showAnswer() {
71         Text answerText = (Text) findComponentById(ResourceTable.Id_text_answer);
72         String answer = selectedSet.toString();
73         answerText.setText(answer);
74     }
75 }
```

打开 Tools>Device Manager,登录账号后,运行远程模拟器查看具体页面的显示效果。单击"全选"按钮,三个选项都会进入选项框;单击"全不选"按钮,选项框会清空,如图 2-24 所示。

图 2-24 全选、全不选效果

2.2.8　Image 组件

Image 是用来显示图片的组件。Image 的共有 XML 属性继承自 Component。Image 的自有 XML 属性如表 2-8 所示。

表 2-8　Image 的自有 XML 属性

属性名称	说明	使用案例
clip_alignment	图片裁剪对齐方式	ohos：clip_alignment＝"left"
image_src	图片资源	ohos：image_src＝"#FFFFFFFF"
scale_mode	图片缩放类型	ohos：scale_mode＝"center"

接下来实现如图 2-25 所示的 Image 效果。

首先在 Project 窗口打开 entry＞src＞main＞resources＞base＞media，拖动所需图片文件添加至 media 文件夹下，以"img1.png"为例，如图 2-26 所示。

layout 目录下的 ability_main.xml 布局文件代码如程序清单 2-29 所示。

图 2-25　Image 页面显示效果

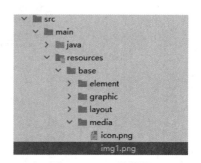

图 2-26　img 位置图

程序清单 2-29　hmos\ch02\07\Checkbox\entry\src\main\resources\base\layout\ability_main.xml

```
1    <?xml version="1.0" encoding="utf-8"?>
2    <DirectionalLayout
3        xmlns:ohos="http://schemas.huawei.com/res/ohos"
```

```
4        ohos:height="match_parent"
5        ohos:width="match_parent"
6        ohos:orientation="vertical">
7        <Image
8            ohos:id="$+id:imageComponent"
9            ohos:height="200vp"
10           ohos:width="200vp"
11           ohos:image_src="$media:img1"
12           ohos:alpha="0.5"
13           ohos:scale_x="1.5"
14           ohos:scale_y="1.5"
15           ohos:top_margin="100fp"
16           ohos:layout_alignment="horizontal_center"/>
17   </DirectionalLayout>
```

2.3 HarmonyOS 布局管理器

2.3.1 DirectionalLayout 布局

DirectionalLayout 布局类似于 Android 中的 LinearLayout 布局，ohos：orientation 的取值有 horizontal 和 vertical 两个，代表横向和纵向布局。DirectionalLayout 用于将一组组件（Component）按照水平或者垂直方向排列，能够方便地对齐布局内的组件。DirectionalLayout 的共有 XML 属性继承自 Component。

DirectionalLayout 的自有 XML 属性如表 2-9 所示。

表 2-9 DirectionalLayout 的自有 XML 属性

属性名称	说明	使用案例
alignment	对齐方式	ohos:alignment="left"
orientation	子布局排列方向	ohos:orientation="horizontal"
total_weight	所有子视图的权重之和	ohos:total_weight="4"

使用 DirectionalLayout 对 3 个文本进行垂直排列，代码如程序清单 2-30 所示。

程序清单 2-30：hmos\ch02\08\DirectionalLayout\entry\src\
main\resources\base\layout\ability_main.xml

```
1    <?xml version="1.0" encoding="utf-8"?>
2    <DirectionalLayout
3        xmlns:ohos="http://schemas.huawei.com/res/ohos"
4        ohos:width="match_parent"
5        ohos:height="match_content"
```

```
 6        ohos:orientation="vertical">
 7        <Text
 8            ohos:width="80vp"
 9            ohos:height="50vp"
10            ohos:background_element="#C4E1E1"
11            ohos:text="Text1"
12            ohos:text_alignment="horizontal_center"
13             ohos:text_size="30vp"/>
14        <Text
15            ohos:width="80vp"
16            ohos:height="50vp"
17            ohos:background_element="#C4E1E1"
18            ohos:text="Text2"
19            ohos:text_alignment="horizontal_center"
20            ohos:text_size="30vp"/>
21        <Text
22            ohos:width="80vp"
23            ohos:height="50vp"
24            ohos:background_element="#C4E1E1"
25            ohos:text="Text3"
26            ohos:text_alignment="horizontal_center"
27            ohos:text_size="30vp"/>
28    </DirectionalLayout>
```

打开 Tools＞Device Manager，登录账号后，运行远程模拟器查看具体页面的显示效果，如图 2-27 所示。

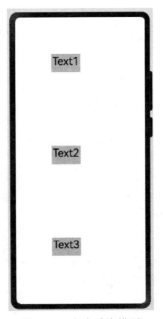

图 2-27　文本垂直排列

使用 DirectionalLayout 对 3 个文本进行水平排列，代码如程序清单 2-31 所示。

程序清单 2-31 hmos\ch02\r8\DirectionalLayout\entry\src\main\resources\base\layout\ability_second.xml

```xml
1   <?xml version="1.0" encoding="utf-8"?>
2   <DirectionalLayout
3       xmlns:ohos="http://schemas.huawei.com/res/ohos"
4       ohos:width="match_parent"
5       ohos:height="match_content"
6       ohos:orientation="horizontal">
7       <Text
8           ohos:width="80vp"
9           ohos:height="50vp"
10          ohos:top_margin="350vp"
11          ohos:left_margin="30vp"
12          ohos:background_element="#C4E1E1"
13          ohos:text="Text1"
14          ohos:text_alignment="horizontal_center"
15          ohos:text_size="30vp"/>
16      <Text
17          ohos:width="80vp"
18          ohos:height="50vp"
19          ohos:top_margin="350vp"
20          ohos:left_margin="30vp"
21          ohos:background_element="#C4E1E1"
22          ohos:text="Text2"
23          ohos:text_alignment="horizontal_center"
24          ohos:text_size="30vp"/>
25      <Text
26          ohos:width="80vp"
27          ohos:height="50vp"
28          ohos:top_margin="350vp"
29          ohos:left_margin="30vp"
30          ohos:background_element="#C4E1E1"
31          ohos:text="Text3"
32          ohos:text_alignment="horizontal_center"
33          ohos:text_size="30vp"/>
34  </DirectionalLayout>
```

打开 Tools＞Device Manager，登录账号后，运行远程模拟器查看具体页面的显示效果，如图 2-28 所示。

权重就是按固定设置的比例分配组件占父组件的位置大小，在水平布局下计算公

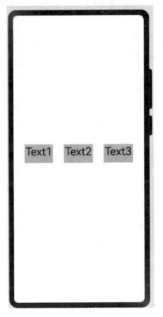

图 2-28　文本水平排列

式为：

父布局可分配宽度＝父布局宽度－所有子组件 width 之和

组件宽度＝组件 weight/所有组件 weight 之和×父布局可分配宽度；在使用过程中，建议使用 width＝0 按比例分配父布局的宽度，这样多个组件都可以平均分配宽度。代码如程序清单 2-32 所示。

程序清单 2-32：hmos\ch02\08\DirectionalLayout\entry\src\main\resources\base\layout\ability_third.xml

```
1   <?xml version="1.0" encoding="utf-8"?>
2   <DirectionalLayout
3       xmlns:ohos="http://schemas.huawei.com/res/ohos"
4       ohos:width="match_parent"
5       ohos:height="match_content"
6       ohos:orientation="horizontal">
7       <Text
8           ohos:width="0vp"
9           ohos:height="50vp"
10          ohos:weight="1"
11          ohos:top_margin="300vp"
12          ohos:left_margin="15vp"
13          ohos:background_element="#C4E1E1"
14          ohos:text="Text1"
15          ohos:text_alignment="horizontal_center"
```

```
16              ohos:text_size="30vp"/>
17      <Text
18              ohos:width="0vp"
19              ohos:height="50vp"
20              ohos:weight="1"
21              ohos:top_margin="300vp"
22              ohos:left_margin="15vp"
23              ohos:background_element="#C4E1E1"
24              ohos:text="Text2"
25              ohos:text_alignment="horizontal_center"
26              ohos:text_size="30vp"/>
27      <Text
28              ohos:width="0vp"
29              ohos:height="50vp"
30              ohos:weight="1"
31              ohos:bottom_margin="300vp"
32              ohos:left_margin="15vp"
33              ohos:background_element="#C4E1E1"
34              ohos:text="Text3"
35              ohos:text_alignment="horizontal_center"
36              ohos:text_size="30vp"/>
37  </DirectionalLayout>
```

打开 Tools>Device Manager,登录账号后,运行远程模拟器查看具体页面的显示效果,如图 2-29 所示。

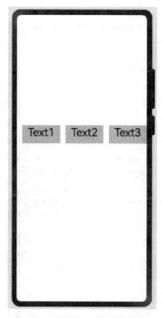

图 2-29　根据权重分配宽度

2.3.2 DependentLayout 布局

DependentLayout 即相对布局,这种布局相对其他布局来说比较常用,因为其十分灵活。相对布局是按照各子元素之间的位置关系完成布局。此外,一个子元素要相对于另一个元素定位,那另一个元素必须先定义。DependentLayout 的共有 XML 属性继承自 Component。DependentLayout 的自有 XML 属性如表 2-10 所示。

表 2-10 DependentLayout 的自有 XML 属性

属性名称	说明	使用案例
below	在某组件下方	ohos:below="＄id:component_id"
above	在某组件上方	ohos:above="＄id:component_id"
left_of	在某组件左侧	ohos:left_of="＄id:component_id"
right_of	在某组件右侧	ohos:right_of="＄id:component_id"
align_parent_left	在父容器左侧	ohos:align_parent_left="true"
align_parent_top	在父容器顶部	ohos:align_parent_top="true"
align_parent_right	在父容器右侧	ohos:align_parent_right="true"
align_parent_bottom	在父容器底部	ohos:align_parent_bottom="true"
align_left	与某组件左侧对齐	ohos:align_left="＄id:component_id"
align_top	与某组件顶部对齐	ohos:align_top="＄id:component_id"
align_right	与某组件右侧对齐	ohos:align_right="＄id:component_id"
align_bottom	与某组件底部对齐	ohos:align_bottom="＄id:component_id"

使用 DependentLayout 实现图 2-30 所示布局,注意"Hello World 3"的定位方法,代码如程序清单 2-33 所示。

图 2-30 DependentLayout 效果图

程序清单 2-23：hmos_ch02_09_DependentLayout\entry\src\main\
resources\base\layout\ability_main.xml

```xml
1  <?xml version="1.0" encoding="utf-8"?>
2  <DependentLayout
3      xmlns:ohos="http://schemas.huawei.com/res/ohos"
4      ohos:height="match_parent"
5      ohos:width="match_parent">
6      <Text
7          ohos:id="$+id:text1"
8          ohos:height="match_content"
9          ohos:width="match_content"
10         ohos:background_element="#FF0000"
11         ohos:layout_alignment="horizontal_center"
12         ohos:text=" Hello World 1 "
13         ohos:align_parent_top = "true"
14         ohos:text_size="50"/>
15     <Text
16         ohos:id="$+id:text2"
17         ohos:height="match_content"
18         ohos:width="match_content"
19         ohos:background_element="#00FF00"
20         ohos:below="$+id:text1"
21         ohos:text=" Hello World 2 "
22         ohos:text_size="50"/>
23     <Text
24         ohos:id="$+id:text3"
25         ohos:height="match_content"
26         ohos:width="match_content"
27         ohos:background_element="#00FF00"
28         ohos:right_of="$+id:text1"
29         ohos:text=" Hello World 4"
30         ohos:text_size="50"/>
31     <Text
32         ohos:id="$+id:text4"
33         ohos:height="match_content"
34         ohos:width="match_content"
35         ohos:background_element="#FF0000"
36         ohos:below="$+id:text3"
37         ohos:right_of="$id:text2"
38         ohos:text=" Hello World 3"
39         ohos:text_size="50"/>
40  </DependentLayout>
```

2.3.3 StackLayout 布局

StackLayout 布局直接在屏幕上开辟出一块空白的区域,当我们往里面添加控件的时候,会默认把它们放到这块区域的左上角,而这种布局方式却没有任何的定位方式,所以它应用的场景并不多;帧布局的大小由控件中最大的子控件决定,如果控件的大小一样大,那么同一时刻只能看到最上面的那个组件。后续添加的控件会覆盖前一个控件。第一个添加到布局中的视图显示在最底层,最后一个视图放在最顶层。上一层的视图会覆盖下一层的视图。

StackLayout 的常用 XML 属性如表 2-11 所示。

表 2-11 StackLayout 的常用 XML 属性

属性名称	说明	使用案例
layout_alignment	对齐方式	ohos：layout_alignment="top|left"

StackLayout 的简单实现:在 layout 目录下新建 ability_main.xml 代码,如程序清单 2-34 所示。

程序清单 2-34：hmos\ch02\10\StackLayout\entry\src\main\resources\base\layout\ability_main.xml

```
1  <?xml version="1.0" encoding="utf-8"?>
2  <StackLayout
3      xmlns:ohos="http://schemas.huawei.com/res/ohos"
4      ohos:height="match_parent"
5      ohos:width="match_parent">
6      <Text
7          ohos:id="$+id:text1"
8          ohos:height="800"
9          ohos:width="800"
10         ohos:background_element="#FF0000"
11         ohos:text_size="50"/>
12     <Text
13         ohos:id="$+id:text2"
14         ohos:height="600"
15  ohos:width="600"
16   ohos:background_element="#00FF00"
17  ohos:text_size="50"/>
18         <Text
19             ohos:id="$+id:text3"
20             ohos:height="400"
21             ohos:width="400"
22             ohos:background_element="#0000FF"
23             ohos:text_size="50"/>
24     </StackLayout>
```

StackLayout 效果图如图 2-31 所示。

2.3.4 TableLayout 布局

TableLayout 布局使用表格的方式划分子组件。如果需要做一个界面,效果类似于九宫格,就需要用这种布局。TableLayout 的共有 XML 属性继承自 Component。

TableLayout 的自有 XML 属性如表 2-12 所示。

表 2-12 TableLayout 的自有 XML 属性

属性名称	说 明	使用案例
column_count	列数	ohos:column_count="3"
row_count	行数	ohos:row_count="2"
orientation	排列方向	ohos:orientation="horizontal"
alignment_type	对齐方式	ohos:alignment_type="align_edges"

使用 TableLayout 实现一个九宫格,效果如图 2-32 所示。

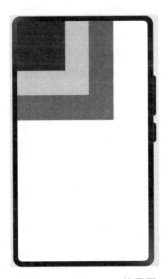

图 2-31 StackLayout 效果图　　图 2-32 TableLayout 九宫格效果图

在 layout 目录下新建 ability_main.xml,代码如程序清单 2-35 所示。

程序清单 2-35：hmos\ch02\11\TableLayout\entry\src\main\resources\base\layout\ability_main.xml

```
1    <?xml version="1.0" encoding="utf-8"?>
2    <TableLayout
3        xmlns:ohos="http://schemas.huawei.com/res/ohos"
4        ohos:height="match_parent"
5        ohos:width="match_parent"
```

```
6          ohos:background_element="blue"
7          ohos:layout_alignment="horizontal_center"
8          ohos:alignment_type="align_contents"
9          ohos:row_count="3"
10         ohos:column_count="3"
11         ohos:orientation="vertical"
12         ohos:padding="8vp">
13         <Text
14             ohos:height="100vp"
15             ohos:width="100vp"
16             ohos:background_element="gray"
17             ohos:margin="8vp"
18             ohos:text="1"
19             ohos:text_alignment="center"
20             ohos:text_size="20fp"/>
21         <Text
22             ohos:height="100vp"
23             ohos:width="100vp"
24             ohos:background_element="gray"
25             ohos:margin="8vp"
26             ohos:text="2"
27             ohos:text_alignment="center"
28             ohos:text_size="20fp"/>
29         <Text
30             ohos:height="100vp"
31             ohos:width="100vp"
32             ohos:background_element="gray"
33             ohos:margin="8vp"
34             ohos:text="3"
35             ohos:text_alignment="center"
36             ohos:text_size="20fp"/>
37         <Text
38             ohos:height="100vp"
39             ohos:width="100vp"
40             ohos:background_element="gray"
41             ohos:margin="8vp"
42             ohos:text="4"
43             ohos:text_alignment="center"
44             ohos:text_size="20fp"/>
45         <Text
46             ohos:height="100vp"
47             ohos:width="100vp"
48             ohos:background_element="gray"
```

```
49          ohos:margin="8vp"
50          ohos:text="5"
51          ohos:text_alignment="center"
52          ohos:text_size="20fp"/>
53      <Text
54          ohos:height="100vp"
55          ohos:width="100vp"
56          ohos:background_element="gray"
57          ohos:margin="8vp"
58          ohos:text="6"
59          ohos:text_alignment="center"
60          ohos:text_size="20fp"/>
61      <Text
62          ohos:height="100vp"
63          ohos:width="100vp"
64          ohos:background_element="gray"
65          ohos:margin="8vp"
66          ohos:text="7"
67          ohos:text_alignment="center"
68          ohos:text_size="20fp"/>
69      <Text
70          ohos:height="100vp"
71          ohos:width="100vp"
72          ohos:background_element="gray"
73          ohos:margin="8vp"
74          ohos:text="8"
75          ohos:text_alignment="center"
76          ohos:text_size="20fp"/>
77      <Text
78          ohos:height="100vp"
79          ohos:width="100vp"
80          ohos:background_element="gray"
81          ohos:margin="8vp"
82          ohos:text="9"
83          ohos:text_alignment="center"
84          ohos:text_size="20fp"/>
85  </TableLayout>
```

2.3.5　PositionLayout 布局

在 PositionLayout 中，子组件通过指定准确的 x、y 坐标值在类似于直角坐标系的屏幕上显示。(0，0) 为左上角，屏幕相当于第四象限；当向下或向右移动时，对应的 x、y 坐标值变大；同时还允许组件之间互相重叠。

PositionLayout 的常用 XML 属性如表 2-13 所示。

表 2-13 PositionLayout 的常用 XML 属性

属性名称	说 明	使用案例
position_x	x 坐标位置	ohos：position_x="120"
position_y	y 坐标位置	ohos：position_y="40"

使用 PositionLayout 实现一个叠加覆盖布局，效果如图 2-33 所示。

图 2-33 PositionLayout 叠加覆盖效果图

在 layout 目录下新建 ability_main.xml，代码如程序清单 2-36 所示。

程序清单 2-36：hmos\ch02\12\PositionLayout\entry\src\main\resources\base\layout\ability_main.xml

```
1   <?xml version="1.0" encoding="utf-8"?>
2   <PositionLayout
3       xmlns:ohos="http://schemas.huawei.com/res/ohos"
4       ohos:height="match_parent"
5       ohos:width="match_parent"
6       ohos:background_element="#66B3FF">
7       <Text
8           ohos:id="$+id:position_text_1"
9           ohos:height="100vp"
10          ohos:width="250vp"
11          ohos:background_element="#FF0000"
12          ohos:position_x="170"
```

```
13          ohos:position_y="40"
14          ohos:text="position1"
15          ohos:text_color="#ffffff"
16          ohos:text_alignment="center"
17          ohos:text_size="20fp"/>
18      <Text
19          ohos:id="$+id:position_text_2"
20          ohos:height="200vp"
21          ohos:width="300vp"
22          ohos:background_element="#00FF00"
23          ohos:position_x="95"
24          ohos:position_y="220"
25          ohos:text="position2"
26          ohos:text_color="#ffffff"
27          ohos:text_alignment="center"
28          ohos:text_size="20fp"/>
29      <Text
30          ohos:id="$+id:position_text_3"
31          ohos:height="200vp"
32          ohos:width="320vp"
33          ohos:background_element="#0000FF"
34          ohos:position_x="65"
35          ohos:position_y="600"
36          ohos:text="position3"
37          ohos:text_color="#ffffff"
38          ohos:text_alignment="center"
39          ohos:text_size="20fp"/>
40  </PositionLayout>
```

2.3.6 AdaptiveBoxLayout 布局

AdaptiveBoxLayout 为自适应盒模式布局，其将整个 UI 划分为多块，主要用于相同级别的多个组件需要在不同屏幕尺寸设备上自动调整列数的应用场景，其中盒子的宽度取决于布局的宽度和每行中盒子的数量，子组件的排列只有在前一行被填满后才会开始在新一行中占位。

使用规则如下。
- 布局中盒子的宽度取决于布局的宽度和每行中盒子的数量，这些属性需要在布局策略中指定。
- 每个盒子的高度由其子组件的高度决定。
- 子组件的排列只有在前一行被填满后才会开始在新一行中占位。
- 每个盒子包含一个子组件。
- 每一行中的所有盒子与高度最高的盒子对齐。

- 布局水平宽度仅支持 match_parent 或固定宽度。
- 某个子组件的高设置为 match_parent 类型,可能导致后续行无法显示。

AdaptiveBoxLayout 布局常用方法如表 2-14 所示。

表 2-14　AdaptiveBoxLayout 布局常用方法

方　　法	功　　能
addAdaptiveRule(int minWidth, int maxWidth, int columns)	添加一个布局规则
removeAdaptiveRule(int minWidth, int maxWidth, int columns)	移除一个布局规则
clearAdaptiveRules()	清除所有布局规则
postLayout()	更新布局

接下来使用代码实现如下两种页面效果。单击 add 按钮,界面会从图 2-34(a)变成图 2-34(b),再单击 remove 按钮,界面会从图 2-34(b)变成图 2-34(a)。

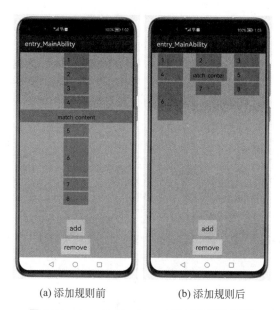

(a) 添加规则前　　　　(b) 添加规则后

图 2-34　AdaptiveBoxLayout 添加规则效果图

在 layout 目录下新建 ability_main.xml,代码如程序清单 2-37 所示。

程序清单 2-37：\hmos\ch2_13_AdaptiveLayout\entry\src\main\resources\base\layout\ability_main.xml

```
1  <?xml version="1.0" encoding="utf-8"?>
2  <DirectionalLayout
3      xmlns:ohos="http://schemas.huawei.com/res/ohos"
4      ohos:height="match_parent"
5      ohos:width="match_parent"
```

```
6       ohos:orientation="vertical"
7       ohos:background_element="#66B3FF">
8       <AdaptiveBoxLayout
9           xmlns:ohos="http://schemas.huawei.com/res/ohos"
10          ohos:height="0vp"
11          ohos:width="match_parent"
12          ohos:weight="1"
13          ohos:id="$+id:adaptive_box_layout">
14          <Text
15              ohos:height="40vp"
16              ohos:width="80vp"
17              ohos:background_element="#0072E3"
18              ohos:margin="2vp"
19              ohos:padding="10vp"
20              ohos:text="1"
21              ohos:text_size="18fp" />
22          <Text
23              ohos:height="40vp"
24              ohos:width="80vp"
25              ohos:background_element="#0072E3"
26              ohos:margin="2vp"
27              ohos:padding="10vp"
28              ohos:text="2"
29              ohos:text_size="18fp" />
30          <Text
31              ohos:height="40vp"
32              ohos:width="80vp"
33              ohos:background_element="#0072E3"
34              ohos:margin="2vp"
35              ohos:padding="10vp"
36              ohos:text="3"
37              ohos:text_size="18fp" />
38          <Text
39              ohos:height="40vp"
40              ohos:width="80vp"
41              ohos:background_element="#0072E3"
42              ohos:margin="2vp"
43              ohos:padding="10vp"
44              ohos:text="4"
45              ohos:text_size="18fp" />
46          <Text
47              ohos:height="40vp"
48              ohos:width="match_parent"
```

```
49            ohos:background_element="#0072E3"
50            ohos:margin="2vp"
51            ohos:padding="10vp"
52            ohos:text="match_content"
53            ohos:text_alignment="horizontal_center"
54            ohos:text_size="18fp" />
55        <Text
56            ohos:height="40vp"
57            ohos:width="80vp"
58            ohos:background_element="#0072E3"
59            ohos:margin="2vp"
60            ohos:padding="10vp"
61            ohos:text="5"
62            ohos:text_size="18fp" />
63        <Text
64            ohos:height="120vp"
65            ohos:width="80vp"
66            ohos:background_element="#0072E3"
67            ohos:margin="2vp"
68            ohos:padding="10vp"
69            ohos:text="6"
70            ohos:text_size="18fp" />
71        <Text
72            ohos:height="40vp"
73            ohos:width="80vp"
74            ohos:background_element="#0072E3"
75            ohos:margin="2vp"
76            ohos:padding="10vp"
77            ohos:text="7"
78            ohos:text_size="18fp" />
79        <Text
80            ohos:height="40vp"
81            ohos:width="80vp"
82            ohos:background_element="#0072E3"
83            ohos:margin="2vp"
84            ohos:padding="10vp"
85            ohos:text="8"
86            ohos:text_size="18fp" />
87    </AdaptiveBoxLayout>
88    <Button
89        ohos:id="$+id:add_btn"
90        ohos:layout_alignment="horizontal_center"
91        ohos:top_margin="10vp"
```

```
92              ohos:padding="10vp"
93              ohos:background_element="#A9CFF0"
94              ohos:height="match_content"
95              ohos:width="match_content"
96              ohos:text_size="22fp"
97              ohos:text="add"/>
98      <Button
99              ohos:id="$+id:remove_btn"
100             ohos:padding="10vp"
101             ohos:top_margin="10vp"
102             ohos:layout_alignment="horizontal_center"
103             ohos:bottom_margin="10vp"
104             ohos:background_element="#D5D5D5"
105             ohos:height="match_content"
106             ohos:width="match_content"
107             ohos:text_size="22fp"
108             ohos:text="remove"/>
109 </DirectionalLayout>
```

通过使用 addAdaptiveRule() 添加一个自适应盒子布局规则，removeAdaptiveRule() 方法用于移除规则，postLayout() 方法用于更新布局。Java 关键代码如程序清单 2-38 所示。

程序清单 2-38：hmos\ch02\13\AdaptiveLayout\entry\src\main\java\com\example\adaptivelayout\slice\MainAbilitySlice.java

```java
1   public class MainAbilitySlice extends AbilitySlice {
2       @Override
3       public void onStart(Intent intent) {
4           super.onStart(intent);
5           super.setUIContent(ResourceTable.Layout_ability_main);
6           AdaptiveBoxLayout adaptiveBoxLayout = (AdaptiveBoxLayout) findComponentById(ResourceTable.Id_adaptive_box_layout);
7           findComponentById(ResourceTable.Id_add_btn).setClickedListener((component-> {
8           //添加规则
9            adaptiveBoxLayout.addAdaptiveRule(100, 3000, 3);
10          //更新布局
11           adaptiveBoxLayout.postLayout();
12      }));
13     findComponentById(ResourceTable.Id_remove_btn).setClickedListener((component-> {
14          //移除规则
15           adaptiveBoxLayout.removeAdaptiveRule(100, 3000, 3);
```

```
16          //更新布局
17          adaptiveBoxLayout.postLayout();
18      }));
19  }
20 }
```

2.4 HarmonyOS 高级界面控件

2.4.1 ListContainer 列表

ListContainer 是鸿蒙应用开发中常用的一种用来呈现连续、多行数据的组件,包含一系列相同类型的列表项,类似于 Android 中的 ListView,其用法和 ListView 的用法极其相似,Android 开发者很容易上手使用。ListContainer 的共有 XML 属性继承自 Component。ListContainer 的自有 XML 属性如表 2-15 所示。

表 2-15 ListContainer 的自有 XML 属性

属性名称	说明	使用案例
shader_color	着色器的颜色	ohos：shader_color="black"
orientation	列表项的排列方向	ohos：orientation="horizontal"
rebound_effect	开启/关闭回弹效果	ohos：rebound_effect="true"

接下来实现如图 2-35 所示的 ListContainer 效果。

图 2-35 ListContainer 效果

在 layout 文件的 ability_main.xml 中声明 ListContainer 控件，代码如程序清单 2-39 所示。

程序清单 2-39：hmos\ch02\14\ListContainer\entry\src\main\resources\base\layout\ability_main.xml

```
1   <?xml version="1.0" encoding="utf-8"?>
2   <DirectionalLayout
3       xmlns:ohos="http://schemas.huawei.com/res/ohos"
4       ohos:width="match_parent"
5       ohos:height="match_parent"
6       ohos:orientation="vertical">
7       <ListContainer
8           ohos:id="$+id:listText"
9           ohos:height="match_parent"
10          ohos:width="match_parent"/>
11  </DirectionalLayout>
```

ListContainer 中的每行也需要代码进行声明，如程序清单 2-40 所示。

程序清单 2-40：hmos\ch02\14\ListContainer\entry\src\main\resources\base\layout\item_sample.xml

```
1   <?xml version="1.0" encoding="utf-8"?>
2   <DirectionalLayout
3       xmlns:ohos="http://schemas.huawei.com/res/ohos"
4       ohos:height="match_content"
5       ohos:width="match_parent"
6       ohos:left_margin="16vp"
7       ohos:right_margin="16vp"
8       ohos:orientation="vertical">
9       <Text
10          ohos:id="$+id:item_index"
11          ohos:height="match_content"
12          ohos:width="match_content"
13          ohos:padding="4vp"
14          ohos:text="Item0"
15          ohos:text_size="20fp"
16          ohos:layout_alignment="center"/>
17  </DirectionalLayout>
```

在 Android 中使用 ListView 会用到适配器 Adapter。适配器是数据和视图之间的桥梁，是连接后端数据和前端显示的接口，在鸿蒙中也有类似的适配器 BaseItemProvider，它作为一个可以把该列表复杂数据简化后并显示给用户的工具，可将 ListContainer 列表数据完整显示出来。使用适配器，该类有多个方法可以实现数据类的交互，方法功能会在代码中介绍,其实现代码如程序清单 2-41 所示。

程序清单 2-41：hmos\ch02\14\ListContainer\entry\src\main\java\com\example\listcontainer\SampleItemProvider.java

```
1   public class SampleItemProvider extends BaseItemProvider {//该自定义鸿蒙适
    //配器需要继承 BaseItemProvider,并重写其中的四个方法
2       private List<String> data;              //需要显示的 List 数据
3       LayoutScatter layoutScatter;//组件转换工具,可以使用该工具在 XML 文件中定
    //义组件,然后调用 parse()方法以使用此文件生成组件对象
4       public SampleItemProvider(Context context, List<String> data) {
5           this.data = data;
6           this.layoutScatter = LayoutScatter.getInstance(context);
7       }
8       @Override
9       public int getCount() {                 //返回填充的表项个数
10          return data.size();
11      }
12      @Override
13      public Object getItem(int i) {          //根据 position 返回对应的数据
14          return null;
15      }
16      @Override
17      public long getItemId(int i) {          //返回某一项的 id
18          return i;
19      }
20      @Override
21  public Component getComponent(int position, Component component,
    ComponentContainer componentContainer) {//根据 position 返回对应的界面组件
22          ViewHolder viewHolder = null;       //缓存
23          if (component == null) {//parse()方法的三个参数分别代表布局文件的 id、父
                                    //组件、布尔值
24        component = layoutScatter.parse(ResourceTable.Layout_item_sample,
    null, false);
25          viewHolder = new ViewHolder((ComponentContainer) component);
26          component.setTag(viewHolder);
27          } else {
28          viewHolder = (ViewHolder) component.getTag();
29          }
30          viewHolder.ItemName.setText(data.get(i));
31          return component;
32      }
33      static class ViewHolder {
34          Text ItemName;
```

```
35        public ViewHolder(ComponentContainer componentContainer) {
36   ItemName = (Text)componentContainer.findComponentById(ResourceTable.Id_
     item_index);
37        }
38      }
39    }
```

其中 ViewHolder 是用来实现 ListContainer 的缓存的，将已经显示在屏幕上的 item 缓存在 ViewHolder 中，下次若再出现，则直接从缓存中读取。parse()方法的第一个参数是布局文件的 id，系统首先会解析布局文件，使用反射将布局里面的组件创建出来，解析完成后就会返回一个 component 对象，第二个参数和第三个参数结合在一起使用，如果第二个参数不为空并且第三个参数为 true，系统就会把刚刚返回的 component 对象放入父组件中，最后将父组件返回。如果第二个参数为空或者第三个参数为 false，系统就不会把 component 对象放入父组件中，仅是将 component 对象返回。

最后需要在 Ability 中给 ListContainer 设置数据，在 Ability 中声明 ListContainer 对象，然后将 ListProvider 对象设置进 ListContainer 中，显示列表数据。实现代码如程序清单 2-42 所示。

程序清单 2-42：hmos\ch02\14\ListContainer\entry\src\main\java\com\example\Test\slice\MainAbilitySlice.java

```
1   public class MainAbilitySlice extends AbilitySlice {
2       @Override
3       protected void onStart(Intent intent) {
4         super.onStart(intent);
5         super.setUIContent(ResourceTable.Layout_ability_main);
6         ListContainer listText = (ListContainer) findComponentById
    (ResourceTable.Id_listText);
7         SampleItemProvider listProvider=new SampleItemProvider(this,
    getList());
8         listText.setItemProvider(listProvider);//设置适配器
9         listText.setItemClickedListener(new ListContainer.
    ItemClickedListener() {
10          @Override
11      public void onItemClicked(ListContainer listContainer, Component
    component, int i, long l) {
12       new ToastDialog(MainAbilitySlice.this).setContentText(String.format("你
          点击了第%d行", i)).show();
13          }
14        });
15      }
```

```
16    private List<String> getList() {
17        List<String> list = new ArrayList<>();
18        for (int i = 0; i < 20; i++) {
19            list.add(String.format("我是第%d行", i));
20        }
21        return list;
22    }
23 }
```

2.4.2　CommonDialog 对话框

CommonDialog 可以在当前的界面弹出一个对话框,这个对话框置顶于所有界面元素之上,用户无法操作除 CommonDialog 外的其他界面内容,能够屏蔽掉其他控件的交互能力,因此 CommonDialog 一般用于提示一些非常重要的内容或警告信息。比如,为了防止用户误删重要内容,在删除前弹出一个确认对话框,接下来通过一个例子实现如图 2-36 所示的对话框。

图 2-36　CommonDialog 实现警告框

CommonDialog 的常用方法包括:CommonDialog(Context context)用于创建一个对话框实例,setTitleText(String text)用于在标题区域中设置标题文本,setContentText (String text)用于设置要在内容区域中显示的文本等一系列方法。Java 实现关键代码片段如程序清单 2-43 所示。

程序清单 2-43：hmos\ch02\15\CommonDialog\entry\src\main\java\com\example\commondialog\MainAbility.java

```java
1   public class MainAbility extends Ability {
2       private Text mtext;
3       @Override
4       public void onStart(Intent intent) {
5           super.onStart(intent);
6           super.setUIContent(ResourceTable.Layout_ability_main);
7   int dialogWidth = (int) (DisplayManager.getInstance().getDefaultDisplay(this).get().getAttributes().width * 0.7);
8           mtext = (Text) findComponentById(ResourceTable.Id_text_helloworld);
9           mtext.setClickedListener(component -> {
10              CommonDialog dialog = new CommonDialog(this);   //新建一个对话框
11              dialog.setTitleText("提示");                      //设置对话框标题
12              dialog.setContentText("是否删除闹钟");             //设置对话框的内容
13              dialog.setSize(dialogWidth, ComponentContainer.LayoutConfig.MATCH_CONTENT);                                     //设置对话框的尺寸
14              dialog.setButton(0, "取消", new IDialog.ClickedListener() {
15                  @Override
16                  public void onClick(IDialog iDialog, int i) {
17                  }
18              });
19              dialog.setButton(1, "删除", new IDialog.ClickedListener() {
20                  @Override
21                  public void onClick(IDialog iDialog, int i) {
22                      dialog.destroy();                        //销毁对话框
23                  }
24              });
25              dialog.show();                                    //展示对话框
26          });
27      }
28  }
```

2.4.3 RadioContainer 单选按钮容器

RadioContainer 是 RadioButton 的容器，在其包裹下的 RadioButton 保证只有一个被选项。其中 RadioButton 用于多选一的操作，需要搭配 RadioContainer 使用，实现单选效果。RadioContainer 的共有 XML 属性继承自 DirectionalLayout，RadioButton 的共有 XML 属性继承自 Text。RadioButton 的自有 XML 属性如表 2-16 所示。

表 2-16 RadioButton 的自有 XML 属性

属性名称	说明	使用案例
marked	当前是否被选中	ohos:marked="true"
text_color_on	选中状态文本颜色	ohos:text_color_on="blue"
text_color_off	未选中状态文本颜色	ohos:text_color_off="black"

接下来实现如图 2-37 所示的单选题效果。

图 2-37 RadioContainer 实现单选题

在 layout 文件的 ability_main.xml 中声明 RadioContainer 控件，代码如程序清单 2-44 所示。

程序清单 2-44：hmos\ch02\16\RadioContainer\entry\src\main\resources\base\layout\ability_main.xml

```
1  <?xml version="1.0" encoding="utf-8"?>
2  <DirectionalLayout
3      xmlns:ohos="http://schemas.huawei.com/res/ohos"
4      ohos:height="match_parent"
5      ohos:width="match_parent"
6      ohos:alignment="horizontal_center"
7      ohos:orientation="vertical"
8      ohos:left_padding="32vp">
9      <Text
```

```
10          ohos:height="match_content"
11          ohos:width="300vp"
12          ohos:top_margin="32vp"
13          ohos:text="RadioContainer 的共有 XML 属性继承自？"
14          ohos:text_size="20fp"
15          ohos:layout_alignment="left"
16          ohos:multiple_lines="true"/>
17      <DirectionalLayout
18          ohos:height="match_content"
19          ohos:width="match_parent"
20          ohos:orientation="horizontal"
21          ohos:top_margin="8vp">
22          <Text
23              ohos:id="$+id:text_checked"
24              ohos:height="match_content"
25              ohos:width="match_content"
26              ohos:text_size="20fp"
27              ohos:left_margin="18vp"
28              ohos:text="[]"
29              ohos:text_color="#FF3333"/>
30      </DirectionalLayout>
31      <RadioContainer
32          ohos:id="$+id:radio_container"
33          ohos:height="match_content"
34          ohos:width="200vp"
35          ohos:layout_alignment="left"
36          ohos:orientation="vertical"
37          ohos:top_margin="16vp"
38          ohos:left_margin="4vp">
39          <RadioButton
40              ohos:id="$+id:radio_button_1"
41              ohos:height="40vp"
42              ohos:width="match_content"
43              ohos:text="A.DirectionalLayout"
44              ohos:text_size="20fp"
45              ohos:text_color_on="#FF3333"/>
46          <RadioButton
47              ohos:id="$+id:radio_button_2"
48              ohos:height="40vp"
49              ohos:width="match_content"
50              ohos:text="B.DependentLayout"
51              ohos:text_size="20fp"
52              ohos:text_color_on="#FF3333"/>
```

```
53          <RadioButton
54              ohos:id="$+id:radio_button_3"
55              ohos:height="40vp"
56              ohos:width="match_content"
57              ohos:text="C.StackLayout"
58              ohos:text_size="20fp"
59              ohos:text_color_on="#FF3333"/>
60          <RadioButton
61              ohos:id="$+id:radio_button_4"
62              ohos:height="40vp"
63              ohos:width="match_content"
64              ohos:text="D.TableLayout"
65              ohos:text_size="20fp"
66              ohos:text_color_on="#FF3333"/>
67      </RadioContainer>
68  </DirectionalLayout>
```

Java 实现关键代码片段如程序清单 2-45 所示。

程序清单 2-45：hmos\ch02\16\RadioContainer\entry\src\main\java\com\text\radiocontainer\slice\MainAbilitySlice.java

```
1   public void onStart(Intent intent) {
2       super.onStart(intent);
3       super.setUIContent(ResourceTable.Layout_ability_main);
4       RadioContainer radioContainer = (RadioContainer) findComponentById(ResourceTable.Id_radio_container);
5       Text answer = (Text) findComponentById(ResourceTable.Id_text_checked);
6       radioContainer.setMarkChangedListener((radioContainer1, index) -> {
7           answer.setText(String.format("[%c]",(char)('A'+index)));
8       });
9   }
```

2.5 本章小结

本章主要讲解了 HarmonyOS 中界面控件的基本知识，应用中所有的用户界面元素都由 Component 和 ComponentContainer 对象构成。Component 是绘制在屏幕上的一个对象，用户能与之交互。ComponentContainer 是一个用于容纳其他 Component 和 ComponentContainer 对象的容器。除此之外，本章还详细介绍了几种最基本的界面控件的功能和常用属性，包括文本显示框、时间选择器和按钮等。为了使这些控件排列美观，还介绍了 HarmonyOS 中几种常见的布局管理器，包括线性布局、表格布局、依赖布局等。其中线性布局方便，需使用的属性较少，但不够灵活；表格布局通过 row_count 和 column _count 设置行、列数；依赖布局则通过参照物确定各个控件的具体位置；层布局中每个控件单独占一层，通过多层叠加形成最终的显示效果；在绝对布局中，子组件通过指定准确的 x、y 坐标值在屏幕上显示。自适应盒子布局提供了在不同屏幕尺寸设备上的自适应布局能力。通常，在一个实例中会用到多种布局，把各种布局结合起来可达到所要的界面效果。

2.6 课后习题

1. 下列 XML 属性（　　）不属于 Text。
   A. ohos:hint   C. ohos:text
   B. ohos:text_font   D. ohos:scale_mode
2. 下列不属于 HarmonyOS 中的布局管理器的是（　　）。
   A. FrameLayout   C. DirectionalLayout
   B. DependentLayout   D. StackLayout
3. 在 DirectionalLayout 中，如果想让一个控件居中显示，可设置该控件的（　　）。
   A. ohos:orientation="center"   C. ohos:center="true"
   B. ohos:alignment="center"   D. ohos:horizontal_center="true"
4. StackLayout 视图以＿＿＿＿方式显示，它会把这些视图默认放到这块区域的角，第一个添加到布局中的视图显示在＿＿＿＿层，最后一个视图被放在＿＿＿＿层。上一层的视图会覆盖下一层的视图。
5. 使用 DependentLayout 或者 TableLayout 实现：界面中包含 4 个 Text 文本，其中 3 个组件分布在各个方向，1 个组件放在屏幕中间。

第 3 章 HarmonyOS 事件处理

课程思政 3

本章要点

- HarmonyOS 基于监听的事件处理
- HarmonyOS 线程管理
- HarmonyOS 线程间通信

本章知识结构图（见图 3-1）

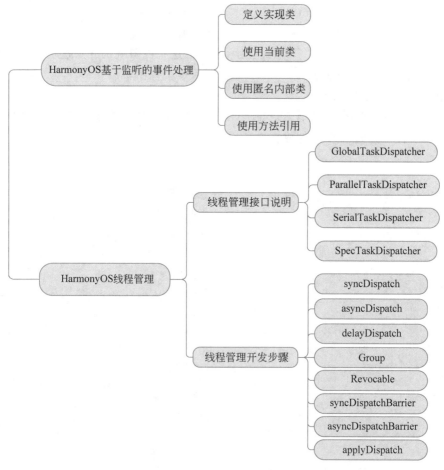

图 3-1　本章知识结构图

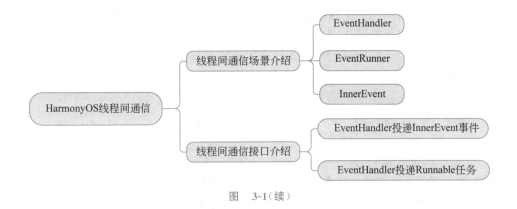

图 3-1（续）

本章示例（见图 3-2）

图 3-2　本章示例图

通过前面两章学习了 HarmonyOS 所提供的一些功能强大的界面控件，这些控件主要用来进行界面的设计与数据的显示，只能实现静态 UI，但是如果用户想与之进行动态交互，获取用户的意图信息，实现更加具体的功能，则还需要相应事件处理的辅助。当用户在程序界面上执行各种操作时，如单击一个按钮，此时一个体验良好的应用程序通常会做出相应的响应，这种响应就是通过事件处理来完成的。在前面章节的学习中，已经使用到了 HarmonyOS 中的一些事件处理，例如单击按钮实现数字加一。

在 HarmonyOS 中，主线程又可以叫作 UI 线程，用户界面属于主线程，默认情况下，所有的操作都在主线程上执行。如果需要执行比较耗时的任务，可创建其他线程（或子线程）来处理，否则主线程会因为耗时操作被阻塞，从而出现异常，所以需要不同的任务分发器来应对不同的操作情况。此外，还需要思考如何才能动态地显示用户界面。本章将介绍通过 EventHandler 消息传递来动态更新界面。

如果在事件处理中需要做一些比较耗时的操作，直接放在主线程中将会阻塞程序的运行,给用户不好的体验,甚至程序会没有响应或强制退出。本章将学习通过同步派发任务以外的方式来处理耗时的操作。

学完本章之后,再结合前面所学知识,读者将可以开发出界面友好、人机交互良好的HarmonyOS应用。

3.1 HarmonyOS 基于监听的事件处理

不管是什么手机应用,都离不开与用户的交互,只有通过用户的操作,才能知道用户的需求,从而实现具体的业务功能,因此,应用中经常需要处理的就是用户的操作,也就是需要为用户的操作提供响应,这种为用户操作提供响应的机制就是事件处理。

HarmonyOS 的基于监听的事件处理模型,与 Java 的 AWT、Swing 的处理方式几乎完全一样,只是相应的事件监听器和事件处理方法名有所不同。在基于监听的事件处理模型中,主要涉及以下 3 类对象。

- Eventsource(事件源):即事件发生的源头,通常为某一控件,如图片、列表等。
- Event(事件):用户具体某一操作的详细描述。事件封装了该操作的相关信息,如果程序需要获得事件源上所发生事件的相关信息,一般通过 Event 对象来取得,例如按键事件按下的是哪个键、触摸事件发生的位置等。
- Eventlistener(事件监听器):负责监听用户在事件源上的操作,并对用户的各种操作做出相应的响应。事件监听器中可包含多个事件处理器,一个事件处理器实际上就是一个事件处理方法。

在基于监听的事件处理中,这 3 类对象又是如何协作的呢? 实际上,基于监听的事件处理是一种委托式事件处理。事件源即普通控件将整个事件处理委托给特定的对象——事件监听器;当该事件源发生指定的事情时,系统自动生成事件对象,并通知所委托的事件监听器,由事件监听器相应的事件处理器处理这个事件。基于监听的事件处理模型如图 3-3 所示。

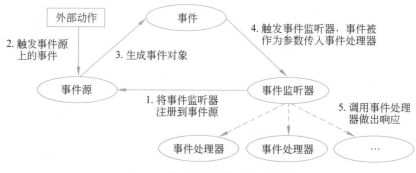

图 3-3　基于监听的事件处理模型

当用户在 HarmonyOS 控件上进行操作时,系统会自动生成事件对象,并将这个事件对象以参数的形式传给注册到事件源上的事件监听器,事件监听器调用相应的事件处理来处理。该过程类似于生活中我们每个人的能力都有限,当碰到一些自己处理不了的事情时,就委托给某个机构处理。我们需要把所遇到的事情和要求描述清楚，这样，其他人才能比较好地解决问题，然后该机构会选派具体的员工来处理这件事。其中，我们自己就是事件源，遇到的事情就是事件，该机构就是事件监听器，具体解决事情的员工就是事件处理器。

单击事件的写法分别为定义实现类、使用当前类作为实现类、使用匿名内部类、使用方法引用四种，接下来通过按钮的单击事件来展示监听事件的处理。

1. 定义实现类

新建一个项目，在 layout 文件夹下打开 ability_main.xml，定义一个按钮，详细代码如程序清单 3-1 所示。

程序清单 3-1：hmos\ch03\01\Listener1\entry\src\main\java\resource\base\layout\ability_main.xml

```xml
1   <?xml version="1.0" encoding="utf-8"?>
2   <DirectionalLayout
3       xmlns:ohos="http://schemas.huawei.com/res/ohos"
4       ohos:height="match_parent"
5       ohos:width="match_parent"
6       ohos:alignment="center"
7       ohos:orientation="vertical">
8       <Button
9           ohos:id="$+id:button1"
10          ohos:height="match_content"
11          ohos:width="match_content"
12          ohos:text="自定义按钮"
13          ohos:padding="15vp"
14          ohos:text_size="100"
15          ohos:background_element="#00ffff"/>
16  </DirectionalLayout>
```

定义实现类实现单击事件，在 MainAbilitySlice 类中设计一个实现 Component.ClickedListener 接口的 MyLister 类，降低代码耦合度，在该类中重写 onClick() 方法从而满足单击事件需求，之后该类的方法可以在主类中的 onStart() 方法中被调用，详细代码如程序清单 3-2 所示。

程序清单 3-2：hmos\ch03\01\Listener1\entry\src\main\java\com\example\listener1\slice\MainAbilitySlice.java

```java
1   public class MainAbilitySlice extends AbilitySlice {
2       @Override
3       public void onStart(Intent intent) {
4           super.onStart(intent);
5           super.setUIContent(ResourceTable.Layout_ability_main);
6           Button button = (Button) findComponentById(ResourceTable.Id_button1);
```

```
7            button.setClickedListener((Component.ClickedListener) new
    MyListener());
8        }
9    class MyListener implements Component.ClickedListener {
10        @Override
11        public void onClick(Component component) {
12            new ToastDialog(getContext())
13                .setText("按钮被单击了")
14                .show();
15        }
16    }
17 }
```

单击按钮屏幕下方会显示一个对话框，效果如图 3-4 所示。

图 3-4　Button 单击示例图

2. 使用当前类作为实现类

使用当前类作为实现类实现单击事件，就是把上个例子中的 MyListener 类与主类合二为一，减少代码量，详细代码如程序清单 3-3 所示，实现效果与图 3-4 相同。

程序清单 3-3：hmos\ch03\01\Listener2\entry\src\main\java\com\
example\listener2\slice\MainAbilitySlice.java

```
1   public class MainAbilitySlice extends AbilitySlice implements Component.
    ClickedListener{
2       @Override
```

```
3      public void onStart(Intent intent) {
4          super.onStart(intent);
5          super.setUIContent(ResourceTable.Layout_ability_main);
6          Button button = (Button) findComponentById(ResourceTable.Id_
    button1);
7          button.setClickedListener(this);
8      }
9      @Override
10     public void onClick(Component component) {
11         new ToastDialog(getContext())
12             .setText("按钮被单击了")
13             .show();
14     }
15 }
```

3. 使用匿名内部类

使用匿名内部类实现单击事件。匿名内部类,就是没有名字的一种嵌套类,使用匿名内部类的方式,可以无须创建新的类,减少代码冗余,详细代码如程序清单 3-4 所示,实现效果与图 3-4 相同。

程序清单 3-4:hmos\ch03\01\Listener3\entry\src\main\java\com\example\listener3\slice\MainAbilitySlice.java

```
1  public class MainAbilitySlice extends AbilitySlice {
2      @Override
3      public void onStart(Intent intent) {
4          super.onStart(intent);
5          super.setUIContent(ResourceTable.Layout_ability_main);
6          Button button = (Button) findComponentById(ResourceTable.Id_
    button1);
7          button.setClickedListener(new Component.ClickedListener() {
8              @Override
9              public void onClick(Component component) {
10                 new ToastDialog(getContext())
11                     .setText("按钮被单击了")
12                     .show();
13             }
14         });
15     }
16 }
```

4. 使用方法引用

使用方法引用实现单击事件,详细代码如程序清单 3-5 所示,实现效果与图 3-4 相同。

程序清单 3-5：hmos\ch03\01\Listener-1\entry\src\main\java\com\example\listener-1\slice\MainAbilitySlice.java

```java
public class MainAbilitySlice extends AbilitySlice{
    @Override
    public void onStart(Intent intent) {
        super.onStart(intent);
        super.setUIContent(ResourceTable.Layout_ability_main);
        Button button = (Button) findComponentById(ResourceTable.Id_button1);
        button.setClickedListener(this::onClick);
    }
    public void onClick(Component component){
        new ToastDialog(getContext())
                .setText("按钮被单击了")
                .show();
    }
}
```

3.2 HarmonyOS 线程管理

进程是应用程序的执行实例，是程序的一次动态执行过程。动态过程是进程本身从产生、发展到最终消亡的过程。线程是比进程更小的执行单位，进程可进一步细化为线程，是一个程序内部的一条执行路径。

多线程是实现并发机制的一种有效手段，指一个进程在执行的过程中产生多个线程，这些线程可以同时存在。一般出现以下 3 种情况需要使用多线程：①程序需要同时执行两个或多个任务时；②程序需要实现一些需要等待的任务时，如用户输入、文件读写操作时；③需要一些后台运行的程序时。

当应用启动时，系统会创建一个主线程。该线程随着应用的动态变化而创建和消亡，是应用的核心线程。主线程负责处理大部分业务，负责应用和 UI 组件发生交互。所以，主线程又称为 UI 线程。因为主线程在任何时候都有较高的响应速度，默认情况下所有操作都在主线程上执行。如果需要执行比较耗时的操作，比如发起一条网络请求时，考虑到网速等原因，服务器未必会立刻响应请求，可创建其他子线程运行这些操作，否则主线程会因为耗时操作被阻塞，从而出现异常，从而影响用户对软件的正常使用。

3.2.1 线程管理接口说明

当应用的业务逻辑复杂时，需要处理多项任务，可能需要开发者创建多个线程来执行多个任务。基于这种情况，代码复杂，难以维护，任务与线程的交互更加繁杂，导致运行效率降低。为了解决这一问题，开发者可以使用 TaskDispatcher 分发不同的任务。

TaskDispatcher 是一个任务分发器,它是 Ability(Ability 在本书 4.1 节有详细介绍)分发任务的基本接口,采用了多线程技术,隐藏任务所在线程的实现细节,TaskDispatcher 获取到执行的任务和方法就可以自动创建合适的线程完成任务。

为保证应用有更好的响应性,需要设计任务的优先级。在 UI 线程上运行的任务默认以高优先级运行,如果某个任务无须等待结果,则可以用低优先级。任务优先级如表 3-1 所示。

表 3-1 任务优先级

优先级	详细描述
HIGH	最高任务优先级,比默认优先级、低优先级的任务有更高的概率执行
DEFAULT	默认任务优先级,比低优先级的任务有更高的概率执行
LOW	低任务优先级,比高优先级、默认优先级的任务有更低的概率执行

TaskDispatcher 具有不同的实现,每种实现对应不同的任务分发器。在分发任务时可以指定任务的优先级,由同一个任务分发器分发出的任务具有相同的优先级。系统提供的任务分发器有 GlobalTaskDispatcher(全局并发任务分发器)、ParallelTaskDispatcher(并发任务分发器)、SerialTaskDispatcher(串行任务分发器)、SpecTaskDispatcher(专有任务分发器)。

- GlobalTaskDispatcher 由 Ability 执行 getGlobalTaskDispatcher(TaskPriority.priority)获取,中文含义为全局并发任务分发器,priority 参数代表任务的优先级,可取值为 HIGH、DEFAULT、LOW,全局并发任务分发器适用于任务没有联系的情况。一个应用只有一个 GlobalTaskDispatcher,它在程序结束时才被销毁,具体代码如程序清单 3-6 所示。

程序清单 3-6

```
1   TaskDispatcher globalTaskDispatcher = getGlobalTaskDispatcher
    (TaskPriority.DEFAULT);
```

- ParallelTaskDispatcher,它由 Ability 执行的 createParallelTaskDispatcher(String name,TaskPriority.priority)创建并返回,中文含义为并发任务分发器,其中 name 参数为任务分发器的名称,priority 参数代表任务的优先级,可取值为 HIGH、DEFAULT、LOW,与 GlobalTaskDispatcher 不同的是,ParallelTaskDispatcher 不具有全局唯一性,可创建多个。开发者在创建或销毁 dispatcher 时,需要持有对应的对象引用,具体代码如程序清单 3-7 所示。

程序清单 3-7

```
1   String dispatcherName = "parallelTaskDispatcher";
2   TaskDispatcher parallelTaskDispatcher = createParallelTaskDispatcher
    (dispatcherName, TaskPriority.DEFAULT);
```

- SerialTaskDispatcher 是由 Ability 执行的 createSerialTaskDispatcher（String name，TaskPriority priority）创建并返回的，中文含义为串行任务分发器，由该分发器分发的所有任务都是按顺序执行的，但是执行这些任务的线程并不是固定的。如果要执行并行任务，应使用 ParallelTaskDispatcher 或者 GlobalTaskDispatcher，而不是创建多个 SerialTaskDispatcher。如果任务之间没有依赖，应使用 GlobalTaskDispatcher 来实现。它的创建和销毁由开发者自己管理，开发者在使用期间需要持有该对象引用，具体代码如程序清单 3-8 所示。

程序清单 3-8

```
1   String dispatcherName = "serialTaskDispatcher";
2   TaskDispatcher serialTaskDispatcher = createSerialTaskDispatcher
    (dispatcherName, TaskPriority.DEFAULT);
```

- SpecTaskDispatcher 的中文含义为专有任务分发器，它是绑定到专有线程上的任务分发器。目前，专有线程是主线程。UITaskDispatcher 和 MainTaskDispatcher 同属 SpecTaskDispatcher。官方更加推荐 UITaskDispatcher。
- UITaskDispatcher：绑定到应用主线程的专有任务分发器，由 getUITaskDispatcher() 创建并返回。由该分发器分发的所有任务都在主线程上按顺序执行，它在应用程序结束时被销毁，具体代码如程序清单 3-9 所示。

程序清单 3-9

```
1   TaskDispatcher uiTaskDispatcher = getUITaskDispatcher();
```

- MainTaskDispatcher：由 Ability 执行 getMainTaskDispatcher() 创建并返回，具体代码如程序清单 3-10 所示。

程序清单 3-10

```
1   TaskDispatcher mainTaskDispatcher= getMainTaskDispatcher();
```

3.2.2 线程管理开发步骤

1. syncDispatch（Runnable runnable）

同步派发任务：派发任务并在当前线程等待任务执行时完成。在返回前，当前线程会被阻塞。同步任务是那些没有被引擎挂起、在主线程上排队执行的任务。只有前一个任务执行完毕，才能执行后一个任务。使用时需实现 Runnable 接口的 run() 方法，在该方法中定义需要执行的任务。

如程序清单 3-11 展示了如何使用 GlobalTaskDispatcher 派发同步任务。

程序清单 3-11：hmos\ch03\02\syncGlobalTaskDispatcher\entry\src\main\java\com\example\syncglobaltaskdispatcher\slice\MainAbilitySlice.java

```java
 1  static final HiLogLabel LABEL = new HiLogLabel(HiLog.LOG_APP, 0x00201, "MY_TAG");
 2  @Override
 3  public void onStart(Intent intent) {
 4      super.onStart(intent);
 5      super.setUIContent(ResourceTable.Layout_ability_main);
 6      TaskDispatcher globalTaskDispatcher = getGlobalTaskDispatcher(TaskPriority.DEFAULT);
 7      globalTaskDispatcher.syncDispatch(new Runnable() {
 8          @Override
 9          public void run() {
10              HiLog.info(LABEL, "同步任务一启动");
11          }
12      });
13      HiLog.info(LABEL, "同步任务一结束");
14      globalTaskDispatcher.syncDispatch(new Runnable() {
15          @Override
16          public void run() {
17              HiLog.info(LABEL, "同步任务二启动");
18          }
19      });
20      HiLog.info(LABEL, "同步任务二结束");
21      globalTaskDispatcher.syncDispatch(new Runnable() {
22          @Override
23          public void run() {
24              HiLog.info(LABEL, "同步任务三启动");
25          }
26      });
27      HiLog.info(LABEL, "同步任务三结束");
28  }
29  //执行结果如下:
30  //同步任务一启动
31  //同步任务一结束
32  //同步任务二启动
33  //同步任务二结束
34  //同步任务三启动
35  //同步任务三结束
```

2. asyncDispatch（Runnable runnable）

异步派发任务：派发任务，并立即返回，返回值是一个可用于取消任务的接口，需要实现 Runnable 接口的 run() 方法，在该方法中定义需要执行的任务。与同步派发任务不同的是，该任务不会阻塞后面代码的执行。

如下代码示例展示了如何使用 GlobalTaskDispatcher 派发异步任务，详细代码如程序清单 3-12 所示。

```
1   public class MainAbilitySlice extends AbilitySlice {
2   static final HiLogLabel LABEL = new HiLogLabel(HiLog.LOG_APP, 0x00201, "MY_TAG");
3       @Override
4       public void onStart(Intent intent) {
5           super.onStart(intent);
6           super.setUIContent(ResourceTable.Layout_ability_main);
7         TaskDispatcher globalTaskDispatcher = getGlobalTaskDispatcher
    (TaskPriority.DEFAULT);
8           globalTaskDispatcher.asyncDispatch(new Runnable() {
9               @Override
10              public void run() {
11                  HiLog.info(LABEL, "async task1 run");
12              }
13          });
14          HiLog.info(LABEL, "after async task1");
15      }
16  }
17  //执行结果如下:
18  //after async task1
19  //async task1 run
```

3. delayDispatch(Runnable runnable, long delay)

异步延迟派发任务：异步执行，函数立即返回，内部会在延时指定时间后将任务派发到相应队列中。该方法的第1个参数表示需执行的任务，第2个参数用于设置执行异步任务延迟的时间。延时时间参数仅代表在这段时间以后任务分发器会将任务加入队列中，任务的实际执行时间可能晚于这个时间。具体比这个数值晚多久，取决于队列及内部线程池的繁忙情况。

如下代码示例展示了如何使用 GlobalTaskDispatcher 延迟派发任务，详细代码如程序清单 3-13 所示。

```
1   public class MainAbilitySlice extends AbilitySlice {
2       static final HiLogLabel LABEL = new HiLogLabel(HiLog.LOG_APP, 0x00201,
    "MY_TAG");
3       @Override
4       public void onStart(Intent intent) {
5           super.onStart(intent);
6           super.setUIContent(ResourceTable.Layout_ability_main);
7           long callTime = System.currentTimeMillis();
```

```
8           long delayTime = 500L;
9           TaskDispatcher globalTaskDispatcher = getGlobalTaskDispatcher
    (TaskPriority.DEFAULT);
10          Revocable revocable = globalTaskDispatcher.delayDispatch(new
    Runnable() {
11              @Override
12              public void run() {
13                  HiLog.info(LABEL, "delayDispatch task1 run");
14                  final long actualDelay = System.currentTimeMillis() - callTime;
15                  HiLog.info(LABEL, "actualDelayTime >= delayTime: %
    {public}b", (actualDelay >= delayTime));
16              }
17          }, delayTime);
18          HiLog.info(LABEL, " delayDispatch task1 finish");
19      }
20  }
21  //执行结果可能如下:
22  //delayDispatch task1 finish
23  //delayDispatch task1 run
24  //actualDelayTime >= delayTime : true
```

4. Group

任务组：表示一组任务，且该组任务之间有一定的联系，由 TaskDispatcher 执行 createDispatchGroup 创建并返回。通过 asyncGroupDispatch（Group group, Runnable runnable）方法将任务加入分组中，通过 groupDispatchNotify（Group group, Runnable runnable）方法可以设置该分组中的所有任务执行完成后再执行指定任务。

如下代码示例展示了任务组的使用方式：将一系列相关联的任务放入一个任务组，执行完任务后关闭应用，详细代码如程序清单 3-14 所示。

程序清单 3-14：hmos\ch03\02\Group\entry\src\main\java\com\example\Group\slice\MainAbilitySlice.java

```
1   public class MainAbilitySlice extends AbilitySlice {
2       static final HiLogLabel LABEL = new HiLogLabel(HiLog.LOG_APP, 0x00201,
    "MY_TAG");
3       @Override
4       public void onStart(Intent intent) {
5           super.onStart(intent);
6           super.setUIContent(ResourceTable.Layout_ability_main);
7           String dispatcherName = "parallelTaskDispatcher";
8           TaskDispatcher dispatcher = createParallelTaskDispatcher
    (dispatcherName, TaskPriority.DEFAULT);
9   //创建一个 group 线程任务组
10          Group group = dispatcher.createDispatchGroup();
```

```
11  //将任务 1 加入线程任务组中,可以返回一个用于取消线程任务的接口
12      dispatcher.asyncGroupDispatch(group, new Runnable() {
13          @Override
14          public void run() {
15              HiLog.info(LABEL, " task1 is running");
16          }
17      });
18  //将与任务 1 相关联的任务 2 加入任务组
19      dispatcher.asyncGroupDispatch(group, new Runnable() {
20          @Override
21          public void run() {
22              HiLog.info(LABEL, " task2 is running");
23          }
24      });
25  //在任务组中的所有任务执行完成后执行指定任务
26      dispatcher.groupDispatchNotify(group, new Runnable() {
27          @Override
28          public void run() {
29  HiLog.info(LABEL, "the close task is running after all tasks in the group are completed");
30          }
31      });
32  //执行结果可能如下:
33  //task1 is running
34  //task2 is running
35  //the close task is running after all tasks in the group are completed
36  //另外一种可能的执行结果:
37  //task2 is running
38  //task1 is running
39  //the close task is running after all tasks in the group are completed
40      }
41  }
```

5. Revocable

取消任务：Revocable 是取消一个异步任务的接口。异步任务包括通过 asyncDispatch、delayDispatch、asyncGroupDispatch 派发的任务。如果任务已经在执行中或已经执行完成，则会返回取消失败的信息。

以下示例展示了如何取消一个异步延时任务，详细代码如程序清单 3-15 所示。

程序清单 3-15：hmos\ch03\02\Revocable\entry\src\main\java\com\example\Revocable\slice\MainAbilitySlice.java

```
1  public class MainAbilitySlice extends AbilitySlice {
2      static final HiLogLabel LABEL = new HiLogLabel(HiLog.LOG_APP, 0x00201, "MY_TAG");
```

```
3       @Override
4       public void onStart(Intent intent) {
5           super.onStart(intent);
6           super.setUIContent(ResourceTable.Layout_ability_main);
7           TaskDispatcher dispatcher = getUITaskDispatcher();
8           Revocable revocable = dispatcher.delayDispatch(new Runnable() {
9               @Override
10              public void run() {
11                  HiLog.info(LABEL, "delay dispatch");
12              }
13          }, 10);
14          boolean revoked = revocable.revoke();
15          HiLog.info(LABEL, "%{public}b", revoked);
16  //一种可能的结果如下：
17  //true
18      }
19  }
```

6. syncDispatchBarrier

同步设置屏障任务：在任务组上设立任务执行屏障，同步等待任务组中的所有任务执行完成，再执行指定任务。

以下示例展示了如何同步设置屏障，详细代码如程序清单 3-16 所示。

程序清单 3-16：hmos\ch03\02\syncDispatchBarrier\entry\src\main\java\com\example\syncDispatchBarrier\slice\MainAbilitySlice.java

```
1   public class MainAbilitySlice extends AbilitySlice {
2     static final HiLogLabel LABEL = new HiLogLabel(HiLog.LOG_APP, 0x00201,
    "MY_TAG");
3       @Override
4       public void onStart(Intent intent) {
5           super.onStart(intent);
6           super.setUIContent(ResourceTable.Layout_ability_main);
7           String dispatcherName = "parallelTaskDispatcher";
8           TaskDispatcher dispatcher = createParallelTaskDispatcher
    (dispatcherName, TaskPriority.DEFAULT);
9   //创建任务组
10          Group group = dispatcher.createDispatchGroup();
11  //将任务加入任务组，返回一个用于取消任务的接口
12          dispatcher.asyncGroupDispatch(group, new Runnable() {
13              @Override
14              public void run() {
15                  HiLog.info(LABEL, "task1 is running");        //1
```

```
16            }
17        });
18        dispatcher.asyncGroupDispatch(group, new Runnable() {
19            @Override
20            public void run() {
21                HiLog.info(LABEL, "task2 is running");        //2
22            }
23        });
24        dispatcher.syncDispatchBarrier(new Runnable() {
25            @Override
26            public void run() {
27                HiLog.info(LABEL, "barrier");                 //3
28            }
29        });
30        HiLog.info(LABEL, "after syncDispatchBarrier");       //4
31 //1和2的执行顺序不定；3和4总是在1和2之后按顺序执行
32 //可能的执行结果：
33 //task1 is running
34 //task2 is running
35 //barrier
36 //after syncDispatchBarrier
37
38 //另外一种执行结果：
39 //task2 is running
40 //task1 is running
41 //barrier
42 //after syncDispatchBarrier
43     }
44 }
```

7. asyncDispatchBarrier

异步设置屏障任务：在任务组上设立任务执行屏障后直接返回，指定任务将在任务组中的所有任务执行完成后再执行。

以下示例展示了如何异步设置屏障，详细代码如程序清单 3-17 所示。

程序清单 3-17：/hmos\ch03\02\asyncDispatchBarrier\entry\src\main\java\com\example\asyncDispatchBarrier\slice\MainAbilitySlice.java

```
1 public class MainAbilitySlice extends AbilitySlice {
2     static final HiLogLabel LABEL = new HiLogLabel(HiLog.LOG_APP, 0x00201, "MY_TAG");
3     @Override
4     public void onStart(Intent intent) {
```

```
5          super.onStart(intent);
6          super.setUIContent(ResourceTable.Layout_ability_main);
7          TaskDispatcher dispatcher = createParallelTaskDispatcher
   ("dispatcherName", TaskPriority.DEFAULT);
8  //创建任务组
9          Group group = dispatcher.createDispatchGroup();
10 //将任务加入任务组,返回一个用于取消任务的接口
11         dispatcher.asyncGroupDispatch(group, new Runnable() {
12             @Override
13             public void run() {
14                 HiLog.info(LABEL, "task1 is running");       //1
15             }
16         });
17         dispatcher.asyncGroupDispatch(group, new Runnable() {
18             @Override
19             public void run() {
20                 HiLog.info(LABEL, "task2 is running");       //2
21             }
22         });
23
24         dispatcher.asyncDispatchBarrier(new Runnable() {
25             @Override
26             public void run() {
27                 HiLog.info(LABEL, "barrier");                //3
28             }
29         });
30         HiLog.info(LABEL, "after asyncDispatchBarrier");     //4
31 //1 和 2 的执行顺序不定,但总在 3 之前执行;4 不需要等待 1、2、3 执行完成
32 //可能的执行结果:
33 //task1 is running
34 //task2 is running
35 //after asyncDispatchBarrier
36 //barrier
37     }
38 }
```

8. applyDispatch

执行多次任务:对指定任务执行多次。以下示例展示了如何执行多次任务,详细代码如程序清单 3-18 所示。

程序清单 3-18 hmos\ch03\02\applyDispatch\entry\src\main\java\com\example\applyDispatch\slice\MainAbilitySlice.java

```
1  public class MainAbilitySlice extends AbilitySlice {
```

```
2        static final HiLogLabel LABEL = new HiLogLabel(HiLog.LOG_APP, 0x00201,
    "MY_TAG");
3        @Override
4        public void onStart(Intent intent) {
5            super.onStart(intent);
6            super.setUIContent(ResourceTable.Layout_ability_main);
7            final int total = 10;
8            final CountDownLatch latch = new CountDownLatch(total);
9            final List<Long> indexList = new ArrayList<>(total);
10           TaskDispatcher dispatcher = getGlobalTaskDispatcher
    (TaskPriority.DEFAULT);
11           //执行任务 total 次
12           dispatcher.applyDispatch((index) -> {
13               indexList.add(index);
14               latch.countDown();
15           }, total);
16           //设置任务超时
17           try {
18               latch.await();
19           } catch (InterruptedException exception) {
20               HiLog.error(LABEL, "latch exception");
21           }
22           HiLog.info(LABEL, "list size matches, %{public}b", (total ==
    indexList.size()));
23           //执行结果：
24           //list size matches, true
25       }
26   }
```

3.3　HarmonyOS 线程间通信

3.3.1　线程间通信场景介绍

在开发过程中，经常需要在当前线程中处理耗时的操作，比如联网读取数据，或者读取本地较大的一个文件，不能把这些操作放在主线程中，因为放在主线程中，界面会出现假死现象，但是又不希望当前的线程受到阻塞，所以会在子线程中执行耗时操作。并且 UI 控件不是线程安全的，所以系统不允许子线程更新 UI，之后提交请求将子线程的数据传递给主线程，让主线程做 UI 更新。此时就可以使用 EventHandler 机制。

EventHandler 是 HarmonyOS 用于处理线程间通信的一种机制，可以在多个线程并发更新 UI 的同时保证线程安全，可以通过 EventRunner 创建新线程，将耗时的操作放到新线程上执行。在不阻塞原来的线程基础下，合理地处理任务。当熟悉了 EventHandler

的原理之后我们知道，EventHandler 不仅能将子线程的数据传递给主线程，它还能实现任意两个线程的数据传递。

EventRunner 是一种事件循环器，循环从该 EventRunner 创建的新线程的事件队列中获取 InnerEvent 事件或者 Runnable 任务。InnerEvent 是 EventHandler 投递的事件。

EventHandler 是一种用户在当前线程上投递 InnerEvent 事件或者 Runnable 任务到异步线程上处理的机制。每一个 EventHandler 和指定的 EventRunner 所创建的新线程绑定，并且该新线程内部有一个事件队列。EventHandler 可以投递指定的 InnerEvent 事件或 Runnable 任务到这个事件队列。EventRunner 从事件队列里循环地取出事件，如果取出的事件是 InnerEvent 事件，将在 EventRunner 所在线程执行 processEvent 回调；如果取出的事件是 Runnable 任务，将在 EventRunner 所在线程执行 Runnable 的 run 回调。一般地，EventHandler 主要有两个作用：①在不同线程间分发和处理 InnerEvent 事件或 Runnable 任务；②延迟处理 InnerEvent 事件或 Runnable 任务。EventHandler 的运作机制如图 3-5 所示。

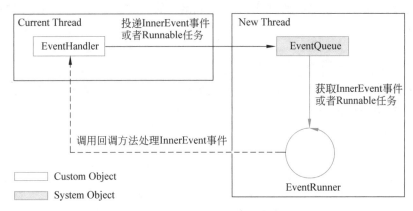

图 3-5　EventHandler 的运作机制

EventHandler 实现线程间通信的主要流程如下。

首先，EventHandler 投递具体的 InnerEvent 事件或者 Runnable 任务到 EventRunner 所创建的线程的事件队列。然后，EventRunner 循环从事件队列中获取 InnerEvent 事件或 Runnable 任务。最后，处理事件或任务，EventRunner 取出事件为 InnerEvent 事件，则触发 EventHandler 的回调方法并触发 EventHandler 的处理方法，在新线程上处理该事件；如果 EventRunner 取出的事件为 Runnable 任务，则 EventRunner 直接在新线程上处理 Runnable 任务。注意：在进行线程间通信的时候，EventHandler 只能和 EventRunner 所创建的线程进行绑定，EventRunner 创建时需要判断是否创建成功，获取的 EventRunner 非空时，才可以使用 EventHandler 绑定 EventRunner。一个 EventHandler 只能同时与一个 EventRunner 绑定，一个 EventRunner 上可以创建多个 EventHandler。

EventHandler 的主要功能是将 InnerEvent 事件或者 Runnable 任务投递到其他线程进行处理，其使用场景包括：

（1）开发者需要将 InnerEvent 事件投递到新的线程，按照优先级和延时进行处理。投递时，EventHandler 的优先级可在 IMMEDIATE、HIGH、LOW、IDLE 中选择，并设置合适的 delayTime。

（2）开发者需要将 Runnable 任务投递到新的线程，并按照优先级和延时进行处理。投递时，EventHandler 的优先级可在 IMMEDIATE、HIGH、LOW、IDLE 中选择，并设置合适的 delayTime。

（3）开发者需要在新创建的线程里投递事件到原线程进行处理。

EventRunner 的工作模式可以分为托管模式和手动模式。两种模式是在调用 EventRunner 的 create()方法时，通过选择不同的参数来实现的，详见 API 参考。默认为托管模式。

托管模式：不需要开发者调用 run()和 stop()方法启动和停止 EventRunner。当 EventRunner 实例化时，系统调用 run()启动 EventRunner；当 EventRunner 不被引用时，系统调用 stop()停止 EventRunner。

手动模式：需要开发者自行调用 EventRunner 的 run()方法和 stop()方法确保线程的启动和停止。

3.3.2 线程间通信接口介绍

EventHandler 的属性 Priority(优先级)介绍：EventRunner 将根据优先级的高低从事件队列中获取事件或者 Runnable 任务进行处理。

EventHandler 的属性如表 3-2 所示。

表 3-2　EventHandler 的属性

属　　性	详　细　描　述
Priority.IMMEDIATE	表示事件被立即投递
Priority.HIGH	表示事件先于 LOW 优先级投递
Priority.LOW	表示事件先于 IDLE 优先级投递，事件的默认优先级是 LOW
Priority.IDLE	表示在没有其他事件的情况下才投递该事件

EventHandler 的主要接口如表 3-3 所示。

表 3-3　EventHandler 的主要接口

接　口　名	详　细　描　述
EventHandler(EventRunner runner)	利用已有的 EventRunner 创建 EventHandler
current()	在 processEvent 回调中获取当前的 EventHandler
processEvent(InnerEvent event)	回调处理事件，由开发者实现
sendEvent(InnerEvent event, long delayTime)	发送一个延时事件到事件队列，优先级为 LOW
sendEvent(InnerEvent event, long delayTime, EventHandler.Priority priority)	发送一个指定优先级的延时事件到事件队列

续表

接口名	详细描述
postSyncTask（Runnable task，EventHandler.Priority priority）	发送一个指定优先级的 Runnable 同步任务到事件队列，延时为 0ms
postTask（Runnable task，long delayTime，EventHandler.Priority priority）	发送一个指定优先级的 Runnable 延时任务到事件队列
sendTimingEvent（InnerEvent event，long taskTime，EventHandler.Priority priority）	发送一个带优先级的事件到队列，在 taskTime 时间执行，如果 taskTime 小于当前时间，则立即执行
postTimingTask（Runnable task，long taskTime，EventHandler.Priority priority）	发送一个带优先级的 Runnable 任务到队列，在 taskTime 时间执行，如果 taskTime 小于当前时间，则立即执行
removeEvent(int eventId，long param，Object object)	删除指定 id、param 和 object 的事件

EventRunner 的主要接口如表 3-4 所示。

表 3-4 EventRunner 的主要接口

接口名	详细描述
create()	创建一个拥有新线程的 EventRunner
create(booleaninNewThread)	创建一个拥有新线程的 EventRunner，参数为 true 时，EventRunner 为托管模式，系统自动管 EventRunner；参数为 false 时，EventRunner 为手动模式
create(String newThreadName)	创建一个拥有新线程的 EventRunner，新线程的名字是 newThreadName
current()	获取当前线程的 EventRunner
run()	EventRunner 为手动模式时，调用该方法启动新的线程
stop()	EventRunner 为手动模式时，调用该方法停止新的线程

InnerEvent 的主要接口如表 3-5 所示。

表 3-5 InnerEvent 的主要接口

接口名	详细描述
drop()	释放一个事件实例
get()	获得一个事件实例
get(inteventId)	获得一个指定的 eventId 的事件实例
get(inteventId，long param)	获得一个指定的 eventId 和 param 的事件实例
get(inteventId，long param，Object object)	获得一个指定的 eventId、param 和 object 的事件实例
get(inteventId，Object object)	获得一个指定的 eventId 和 object 的事件实例
PacMap getPacMap()	获取 PacMap，如果没有，就新建一个

续表

接口名	详细描述
RunnablegetTask()	获取 Runnable 任务
PacMap peekPacMap()	获取 PacMap
voidsetPacMap(PacMap pacMap)	设置 PacMap

InnerEvent 的属性如表 3-6 所示。

表 3-6　InnerEvent 的属性

属性	详细描述
eventId	事件的 ID，由开发者定义，用来辨别事件
object	事件携带的 Object 信息
param	事件携带的 long 型数据

EventHandler 投递 InnerEvent 事件，并按照优先级和延时进行处理，开发步骤如下。

（1）创建 EventHandler 的子类，在子类中重写实现方法 processEvent() 来回调处理事件，详细代码如程序清单 3-19 所示。

程序清单 3-19：hmos\ch03\03\InnerEvent\entry\src\main\java\com\example\InnerEvent\slice\MainAbilitySlice.java

```
1   static final HiLogLabel LABEL = new HiLogLabel(HiLog.LOG_APP, 0x00201, "MY_TAG");
2   private static final int EVENT_MESSAGE_NORMAL = 0x1000001;
3   private static final int EVENT_MESSAGE_DELAY = 0x1000002;
4   private static final int DELAY_TIME = 1000;
5   private class TestEventHandler extends EventHandler {
6       private TestEventHandler(EventRunner runner) {
7           super(runner);
8       }
9       @Override
10      public void processEvent(InnerEvent event) {
11          switch (event.eventId) {
12              case EVENT_MESSAGE_NORMAL:
13                  HiLog.info(LABEL, "Received an innerEvent message");
14                  break;
15              case EVENT_MESSAGE_DELAY:
16                  HiLog.info(LABEL, "Received an innerEvent delay message");
17                  break;
18              default:
19                  break;
20          }
```

```
21        }
22   }
23   //执行结果
24   //Received an innerEvent message
25   //Received an innerEvent delay message
```

（2）创建 EventRunner。创建一个拥有新线程的 EventRunner，以托管模式为例，新线程的名字是 TestRunner，详细代码如程序清单 3-20 所示。

程序清单 3-20：hmos\ch03\03\InnerEvent\entry\src\main\java\com\example\InnerEvent\slice\MainAbilitySlice.java

```
eventRunner = EventRunner.create("TestRunner");
```

（3）创建 EventHandler 子类的实例，详细代码如程序清单 3-21 所示。

程序清单 3-21：hmos\ch03\03\InnerEvent\entry\src\main\java\com\example\InnerEvent\slice\MainAbilitySlice.java

```
TestEventHandler handler = new TestEventHandler(eventRunner);
```

（4）获取 InnerEvent 事件，详细代码如程序清单 3-22 所示。

程序清单 3-22：hmos\ch03\03\InnerEvent\entry\src\main\java\com\example\InnerEvent\slice\MainAbilitySlice.java

```
1   long param = 0L;
2   InnerEvent normalInnerEvent = InnerEvent.get(EVENT_MESSAGE_NORMAL, param, null);
3   InnerEvent delayInnerEvent = InnerEvent.get(EVENT_MESSAGE_DELAY, param, null);
```

（5）投递事件，投递的优先级以 IMMEDIATE 为例，延时选择 0ms 和 1ms，详细代码如程序清单 3-23 所示。

程序清单 3-23：hmos\ch03\03\InnerEvent\entry\src\main\java\com\example\InnerEvent\slice\MainAbilitySlice.java

```
1   handler.sendEvent(normalInnerEvent, EventHandler.Priority.IMMEDIATE);
2   handler.sendEvent(delayInnerEvent, DELAY_TIME, EventHandler.Priority.IMMEDIATE);
```

（6）如果为手动模式，需要启动和停止 EventRunner，详细代码如程序清单 3-24 所示。

程序清单 3-24：hmos\ch03\03\InnerEvent\entry\src\main\java\com\
example\InnerEvent\slice\MainAbilitySlice.java

```
1    runner.run();
2    runner.stop();
```

EventHandler 投递 Runnable 任务，按照优先级和延时进行处理，开发步骤如下。

（1）创建 EventHandler 的子类。创建 EventRunner，并创建 EventHandler 子类的实例，步骤与 EventHandler 投递 InnerEvent 场景的步骤（1）～（3）相同。

（2）创建 Runnable 任务并且投递 Runnable 任务，投递的优先级以 IMMEDIATE 为例，延时选择 0ms 和 1ms，详细代码如程序清单 3-25 所示。

程序清单 3-25：hmos\ch03\03\Runnable\entry\src\main\java\
com\example\Runnable\slice\MainAbilitySlice.java

```
1    private void postRunnableTask(Component component) {
2        stringBuffer = new StringBuffer();
3        Runnable normalTask = new Runnable() {
4            @Override
5            public void run() {
6                HiLog.info(LABEL, "Post runnableTask1 ");
7            }
8        };
9        Runnable delayTask = new Runnable() {
10           @Override
11           public void run() {
12               HiLog.info(LABEL, "Post runnableTask2 ");
13           }
14       };
15       handler.postTask(normalTask, EventHandler.Priority.IMMEDIATE);
16       handler.postTask(delayTask, DELAY_TIME, EventHandler.Priority.IMMEDIATE);
17   }
```

接下来通过一个完整的代码示例实现 EventHandler 投递 InnerEvent 事件和投递 Runnable 任务，界面详细代码如程序清单 3-26 所示。

程序清单 3-26：hmos\ch03\03\EventHandler\entry\src\main\resources\base\layout\ability_main.xml

```
1    <?xml version="1.0" encoding="utf-8"?>
2    <DirectionalLayout
3        xmlns:ohos="http://schemas.huawei.com/res/ohos"
4        ohos:height="match_parent"
5        ohos:width="match_parent"
```

```
6        ohos:alignment="horizontal_center"
7        ohos:orientation="vertical"
8        ohos:padding="5vp">
9        <Button
10           ohos:id="$+id:send_event_button"
11           ohos:height="match_content"
12           ohos:width="300vp"
13           ohos:background_element="$graphic:button_bg"
14           ohos:padding="5vp"
15           ohos:text="Send InnerEvent"
16           ohos:text_alignment="center"
17           ohos:text_size="35fp"
18           ohos:top_margin="80vp"/>
19       <Button
20           ohos:id="$+id:post_task_button"
21           ohos:height="match_content"
22           ohos:width="300vp"
23           ohos:background_element="$graphic:button_bg"
24           ohos:padding="5vp"
25           ohos:text="Post Runnable Task"
26           ohos:text_alignment="center"
27           ohos:text_size="35fp"
28           ohos:top_margin="150vp"/>
29   </DirectionalLayout>
```

按钮背景设置代码 Button_bg.xml 如程序清单 3-27 所示。

```
1   <?xml version="1.0" encoding="utf-8"?>
2   <shape xmlns:ohos="http://schemas.huawei.com/res/ohos"
3       ohos:shape="rectangle">
4       <solid
5           ohos:color="#007CFD"/>
6       <corners
7           Ohos:radius="10"/>
8   </shape>
```

单击按钮投递任务界面如图 3-6 所示。

示例完整代码 MainAbilitySlice.java 如程序清单 3-28 所示。

图 3-6 单击按钮投递任务界面

程序清单 3-28：hmos\ch03\03\EventHandlerDemo\entry\src\main\java\
com\example\eventhandlerdemo\slice\MainAbilitySlice.java

```
1   public class MainAbilitySlice extends AbilitySlice {
2       private static final int EVENT_MESSAGE_NORMAL = 0x1000001;
3       private static final int EVENT_MESSAGE_DELAY = 0x1000002;
4       private static final int EVENT_MESSAGE_CROSS_THREAD = 0x1000003;
5       private static final int DELAY_TIME = 1000;
6       static final HiLogLabel LABEL = new HiLogLabel(HiLog.LOG_APP, 0x00201,
    "MY_TAG");
7       private Text resultText;
8       private StringBuffer stringBuffer = new StringBuffer();
9       private TestEventHandler handler;
10      private EventRunner eventRunner;
11      @Override
12      public void onStart(Intent intent) {
13          super.onStart(intent);
14          super.setUIContent(ResourceTable.Layout_main_ability_slice);
15          initHandler();
16          initComponents();
17      }
18          private void initHandler() {
19              eventRunner = EventRunner.create("TestRunner");
```

```
20            handler = new TestEventHandler(eventRunner);
21        }
22        private void initComponents() {
23            Component sendEventButton = findComponentById(ResourceTable.
    Id_send_event_button);
24            Component postTaskButton = findComponentById(ResourceTable.
    Id_post_task_button);
25            sendEventButton.setClickedListener(this::sendInnerEvent);
26            postTaskButton.setClickedListener(this::postRunnableTask);
27        }
28        private void sendInnerEvent(Component component) {
29            stringBuffer = new StringBuffer();
30            long param = 0L;
31            Object object = null;
32  InnerEvent normalInnerEvent=InnerEvent.get(EVENT_MESSAGE_NORMAL, param,
    object);
33        InnerEvent delayInnerEvent = InnerEvent.get(EVENT_MESSAGE_DELAY,
    param, object);
34        handler.sendEvent(normalInnerEvent,EventHandler.Priority.
    IMMEDIATE);
35        handler.sendEvent(delayInnerEvent, DELAY_TIME, EventHandler.
    Priority.IMMEDIATE);
36        }
37        private void postRunnableTask(Component component) {
38            stringBuffer = new StringBuffer();
39            Runnable normalTask = new Runnable() {
40               @Override
41               public void run() {
42                   HiLog.info(LABEL, "Post runnableTask1 ");
43               }
44            };
45            Runnable delayTask = new Runnable() {
46               @Override
47               public void run() {
48                   HiLog.info(LABEL, "Post runnableTask2 ");
49               }
50            };
51            handler.postTask(normalTask, EventHandler.Priority.IMMEDIATE);
52            handler.postTask(delayTask, DELAY_TIME, EventHandler.
    Priority.IMMEDIATE);
53        }
54        private class TestEventHandler extends EventHandler {
55            private TestEventHandler(EventRunner runner) {
```

```
56                super(runner);
57            }
58            @Override
59            public void processEvent(InnerEvent event) {
60                switch (event.eventId) {
61                    case EVENT_MESSAGE_NORMAL:
62              HiLog.info(LABEL, "Received an innerEvent message");
63                        break;
64                    case EVENT_MESSAGE_DELAY:
65              HiLog.info(LABEL, "Received an innerEvent delay message");
66                        break;
67                    default:
68                        break;
69                }
70            }
71        }
72  }
73  //分别单击两个按钮,HiLog 日志输出情况如下所示。
74  //Received an innerEvent message
75  //Received an innerEvent delay message
76  //Post runnableTask1
77  //Post runnableTask2
```

接下来再通过使用 setClickedListener()、EventHandler、TaskDispatcher 实现一个时钟案例来巩固本章所学的事件监听与线程通信知识,单击 Start 按钮会实时显示当前时间,单击 end 按钮会返回原始界面,显示效果如图 3-7 所示。

(a) 单击Start按钮显示时间　　(b) 单击end按钮返回

图 3-7　时间显示效果图

界面的详细代码如程序清单 3-29 所示。

程序清单 3-29：hmos\ch03\04\clockDemo\entry\src\main\java\resource\base\layout\ability_main.xml

```xml
1   <?xml version="1.0" encoding="utf-8"?>
2   <DirectionalLayout
3       xmlns:ohos="http://schemas.huawei.com/res/ohos"
4       ohos:height="match_parent"
5       ohos:width="match_parent"
6       ohos:alignment="center"
7       ohos:orientation="vertical">
8       <Text
9           ohos:id="$+id:text"
10          ohos:height="match_content"
11          ohos:width="match_content"
12          ohos:background_element="$graphic:background_ability_main"
13          ohos:layout_alignment="horizontal_center"
14          ohos:text="Now time"
15          ohos:text_size="60fp"
16          ohos:top_padding="20vp"/>
17      <Button
18          ohos:id="$+id:btn"
19          ohos:height="match_content"
20          ohos:width="match_content"
21          ohos:text="Start"
22          ohos:text_size="60fp"
23          ohos:top_margin="200vp"
24          ohos:padding="15vp"
25          ohos:background_element="$graphic:button_element"/>
26  </DirectionalLayout>
```

Button 按钮背景代码如程序清单 3-30 所示。

程序清单 3-30：hmos\ch03\04\clockDemo\entry\src\main\java\resource\base\graphic\background_button.xml

```xml
1   <?xml version="1.0" encoding="utf-8"?>
2   <shape xmlns:ohos="http://schemas.huawei.com/res/ohos"
3          ohos:shape="rectangle">
4       <corners
5           ohos:radius="10"/>
6       <solid
7           ohos:color="00ffff"/>
8   </shape>
```

示例完整代码 MainAbilitySlice.java 如程序清单 3-31 所示。

程序清单 3-31：hmos\ch03\04\clockDemo\entry\src\main\java\com\example\clockDemo\slice\MainAbilitySlice.java

```java
1     public class MainAbilitySlice extends AbilitySlice {
2         boolean stop;
3
4         @Override
5         public void onStart(Intent intent) {
6             super.onStart(intent);
7             super.setUIContent(ResourceTable.Layout_ability_main);
8             TaskDispatcher ClockTaskDispatcher = createParallelTaskDispatcher("clock", TaskPriority.DEFAULT);
9             Text dateText = (Text) findComponentById(ResourceTable.Id_text);
10            EventRunner eventRunner = EventRunner.getMainEventRunner();
11            EventHandler eventHandler = new ClockEventHandler(eventRunner, dateText);
12            Button btnClick = (Button) findComponentById(ResourceTable.Id_btn);
13            if (btnClick != null) {
14                btnClick.setClickedListener(new Component.ClickedListener() {
15                    @Override
16                    public void onClick(Component component) {
17                        String btnName = btnClick.getText();
18                        Text dateText = (Text) findComponentById(ResourceTable.Id_text);
19                        if ("Start".equals(btnName)) {
20                            btnClick.setText("end");
21                            stop = false;
22                            ClockTaskDispatcher.asyncDispatch(new Runnable() {
23                                @Override
24                                public void run() {
25                                    while (!stop) {
26                                        eventHandler.sendEvent(1);
27                                        try {
28                                            Thread.sleep(1000);
29                                        } catch (InterruptedException e) {
30                                            e.printStackTrace();
31                                        }
32                                    }
33                                }
34                            });
35                        } else {
```

```
36                        dateText.setText("Now time");
37                        btnClick.setText("Start");
38                        stop = true;
39                    }
40                }
41            });
42    }
43  }
44
45    class ClockEventHandler extends EventHandler {
46        private Text dateText;
47
48        public ClockEventHandler(EventRunner runner) throws
    IllegalArgumentException {
49            super(runner);
50        }
51
52        public ClockEventHandler(EventRunner runner, Text dateText)
    throws IllegalArgumentException {
53            this(runner);
54            this.dateText = dateText;
55        }
56
57        protected void processEvent(InnerEvent event) {
58            super.processEvent(event);
59            Date now = Calendar.getInstance().getTime();
60            SimpleDateFormat clockFormat = new SimpleDateFormat("HH:
    mm:ss");
61            String dateTime = clockFormat.format(now);
62            this.dateText.setText(dateTime);
63        }
64    }
65  }
```

3.4 本章小结

本章是对前面两章的补充，图形界面编程需要与事件处理相结合，设计了界面友好的应用后，必须为界面上的相应控件提供响应，从而当用户操作时，能执行相应的功能，这种响应动作就是由事件处理来完成的。本章的重点就是掌握 HarmonyOS 的事件处理机制。需要注意的是，HarmonyOS 不允许在子线程中更新主线程的界面控件。因此，通过 EventHandler 消息处理机制时，如果需要子线程更改界面显示，子线程就向主线程发送一条消息,主线程接收到消息后,自己对界面显示进行修改。

3.5 课后习题

1. 下列哪个选项不属于线程的优先级?(　　)
   A. HIGH　　　　B. NORMAL　　　　C. DEFAULT　　　　D. LOW
2. 单击事件的写法具有多种实现方式,主要为_____、_____、_____、_____ 4 种。
3. TaskDispatcher 具有多种实现方式,每种实现对应不同的任务分发器,分别为_____、_____、_____、_____ 4 种。
4. 简述什么是 EventRunner,什么是 EventHandler。
5. 简述托管模式和手动模式的区别。

第 4 章 Ability 与 Intent

本章要点

- 理解 Ability 的功能和作用
- 创建和配置 Ability
- Ability 的生命周期
- Page Ability 间的数据传输
- Service Ability 的启动与停止、连接与断开
- Data Ability 相关操作类

本章知识结构图（见图 4-1）

图 4-1　本章知识结构图

本章示例（见图 4-2）

图 4-2　本章示例图

　　HarmonyOS 应用通常由一个或多个组件组成，其中 Ability 是最基础也是最常见的组件，Ability 有 3 种不同的模板，前面所学的程序通常只包含一个 Page Ability 以及一个 AbilitySlice，且并未涉及另外两种模板的使用。本章将详细讲解 Ability 以及其 3 种模板的相关知识，包括 Ability 的创建、配置、启动、停止、数据传输以及它们的生命周期。

　　一个应用程序往往由多个 Page Ability、AbilitySlice 或其他组件组成，那么各 Page Ability 或 AbilitySlice 间是如何交互或者通信的呢？HarmonyOS 中与 Android 中类似，均通过 Intent 对象完成这一功能。本章将讲解 Intent 对象是如何封装组件间的交互的。一个应用通常还涉及 Service Ability 以及 Data Ability 的使用，Service Ability 通常用来处理一些耗时操作，Data Ability 则用来帮助应用管理其自身和其他应用存储数据的访问，本章将详细讲解这两种模板的作用和基本用法。

　　通过本章的学习，读者可以实现 Page Ability 或 AbilitySlice 间数据的传输，以及通过 Service Ability 执行后台任务，完成如用户注册等功能。

4.1　Ability 介绍

　　Ability 是应用所具备能力的抽象，也是应用程序的重要组成部分，一个应用程序可以包含一个或多个 Ability。HarmonyOS 应用程序的 Ability 主要分为两种类型：FA (Feature Ability) 和 PA (Particle Ability)。每种类型均为开发者提供了不同的模板，以便实现不同的业务功能。

- Page 模板：FA 唯一支持的模板，用于提供与用户交互的能力。一个 Page 实例可以包含一个或多个相关页面，每个页面需用一个 AbilitySlice 实例表示。

- Service 模板：PA 支持的模板，用于提供后台运行任务的能力，也用于开发始终在后台运行或与其他功能连接的服务。
- Data 模板：PA 支持的模板，用于对外部提供统一的数据访问抽象。它提供了用于插入、删除、查询和更新数据以及打开文件等方法。

4.1.1 创建一个 Ability

创建一个 Ability，需要在 HUAWEI DevEco Studio 的 Project 窗口中当前工程的主目录（entry＞src＞main＞java＞com.xxx.xxx）上右击，从弹出的快捷菜单中选择 New＞Ability，这里有 3 种 Ability 模板可供选择，分别为 Empty Page Ability（Java）、Empty Data Ability 以及 Empty Service Ability，如图 4-3 所示。

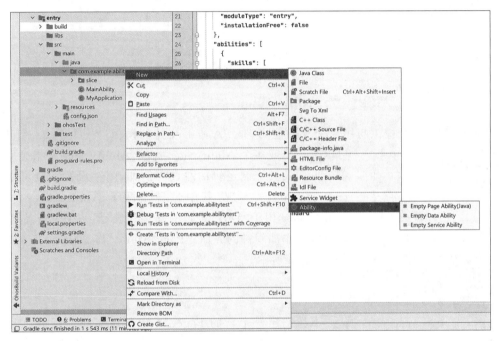

图 4-3　创建 Ability

4.1.2　Ability 的配置

创建好 Ability 后，HUAWEI DevEco Studio 会自动在配置文件（config.json）中注册该 Ability，type 属性的值表示不同的模板，其中 type 的值为 page 时表示该 Ability 使用的是 Page 模板，值为 service 时表示该 Ability 使用的是 Service 模板，值为 data 时表示该 Ability 使用的是 Data 模板，具体代码如程序清单 4-1 所示。

程序清单 4-1：hmos\ch04\01\AbilityTest\entry\src\main\config.json

```
1  {
2      "module": {
```

```
3          ...
4          "abilities": [
5              {
6                  ...
7                  "type": "page"
8                  ...
9              }
10         ]
11         ...
12     }
13     ...
14 }
```

4.2　Page Ability

4.2.1　Page Ability 与 AbilitySlice

Page 模板用于提供与用户交互的能力，是 FA 唯一支持的模板。一个 Page Ability 可以由一个或多个 AbilitySlice 构成，AbilitySlice 是指应用的单个页面及其控制逻辑的总和。例如，闹钟设置功能可以通过一个 Page 来实现，其中包含了两个 AbilitySlice：一个 AbilitySlice 用于展示已设置的闹钟列表；另一个 AbilitySlice 用于展示添加闹钟的界面。Page Ability 与 AbilitySlice 的关系如图 4-4 所示。

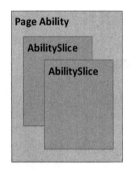

图 4-4　Page Ability 与 AbilitySlice 的关系

4.2.2　创建 Page Ability

下面创建一个新的 Page Ability，在当前工程的主目录（entry＞src＞main＞java＞com.xxx.xxx）上右击，从弹出的快捷菜单中选择 New＞Ability＞Empty Page Ability(Java)，如图 4-5 所示。

然后会进入如图 4-6 所示的界面。

其中，Page Name 里填写需要自定义的 Page Ability 的名称，Package name 里填写新建的 Page Ability 需要放在哪个包下，如果填写的包名不存在，HUAWEI DevEco Studio 会自动创建该包。Layout Name 里填写当前 Page Ability 默认展示的布局文件的名称。均填写完成后单击 Finish 按钮，DevEco Studio 会创建一个 Page Ability，并且在 slice 包下会自动创建一个 AbilitySlice，如图 4-7 所示。

创建完 Page Ability 的同时，DevEco Studio 也会在配置文件（config.json）中注册该 Page Ability，可以注意到其 type 的值为 page，如图 4-8 所示。

第 4 章　Ability 与 Intent

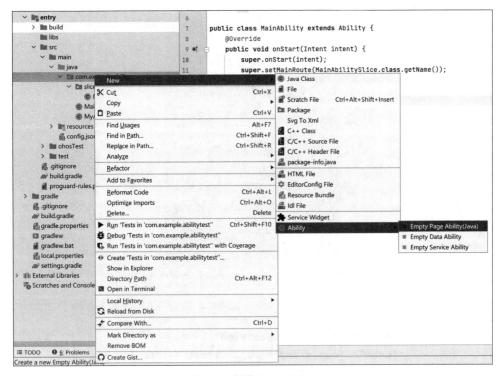

图 4-5　创建 Page Ability

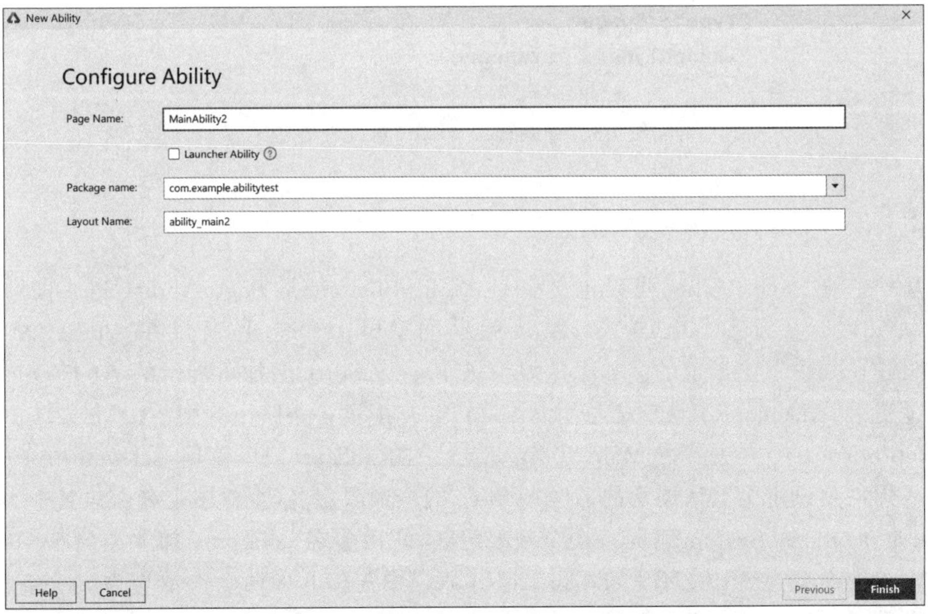

图 4-6　新建 Page Ability 的相关信息

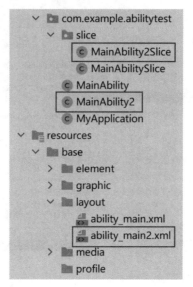

图 4-7 创建 Page Ability 的相关文件

```
{
  "orientation": "unspecified",
  "name": "com.example.abilitytest.MainAbility2",
  "icon": "$media:icon",
  "description": Java_Empty Ability,
  "label": entry_MainAbility2,
  "type": "page",
  "launchType": "standard"
}
```

图 4-8 新建 Page Ability 的配置信息

4.2.3 AbilitySlice 路由配置

虽然一个 Page Ability 可以包含多个 AbilitySlice，但是 Page Ability 进入前台时界面默认只展示一个 AbilitySlice。默认展示的 AbilitySlice 是通过在 Ability 中使用 setMainRoute()方法指定的。如果需要给该 Page Ability 增加可展示的 AbilitySlice，可以通过使用 addActionRoute()方法为此 AbilitySlice 配置一条路由规则。setMainRoute()和 addActionRoute()方法需在 MainAbility.java 中的 onStart()里使用，setMainRoute()中需传入你需要设定为默认展示的 AbilitySlice 的全额限定名(全额限定名由包名＋类名组成，例如 java.lang.System 就是一个全额限定名，它由包名 java.lang 加上类名 System 组成)，该参数类型为 String 型。addActionRoute()中需传入的两个参数均为 String 型，第一个参数是开发者自定义的动作名，第二个参数为需要设定路由的 AbilitySlice 的全额限定名。setMainRoute()方法与 addActionRoute()方法的使用示例如程序清单 4-2 所示。

程序清单 4-2：hmos\ch04\01\AbilityTest\entry\src\main\java\com\example\abilitytest\MainAbility.java

```
1  public class MainAbility extends Ability {
2      @Override
3      public void onStart(Intent intent) {
4          super.onStart(intent);
5          //设置主 AbilitySlice
6          setMainRoute(MainSlice.class.getName());
7          //设置 ActionRoute
8          addActionRoute("action.second", SecondSlice.class.getName());
9      }
10 }
```

addActionRoute()方法中使用的动作命名，需要在应用配置文件（config.json）中进行注册，详细代码如程序清单 4-3 所示。

程序清单 4-3：hmos\ch04\01\AbilityTest\entry\src\main\config.json

```
1  {
2      "module": {
3          "abilities": [
4              {
5                  "skills":[
6                      {
7                          "actions":[
8                              ...
9                              "action.second"
10                         ]
11                     }
12                 ]
13                 ...
14             }
15         ]
16         ...
17     }
18     ...
19 }
```

4.2.4　Page Ability 生命周期

Page Ability 主要负责页面交互，类似于 Android 中的 Activity。Android 中的

Activity 是有生命周期的,同样,Page Ability 也有其自己的生命周期。系统管理或用户操作等行为均会引起 Page Ability 实例在其生命周期的不同状态之间相互转换。Page Ability 能通过 Ability 类提供的回调机制及时感知外界变化,从而正确应对。Page Ability 生命周期的不同状态转换及其对应的回调如图 4-9 所示。

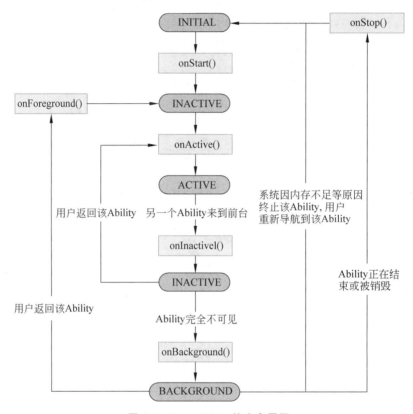

图 4-9 Page Ability 的生命周期

- onStart():当系统首次创建 Page Ability 实例时,触发该回调,这和 Android 中的 onCreate()类似。对于一个 Page Ability 实例,该回调在其生命周期过程中仅触发一次,Page Ability 在该逻辑后将进入 INACTIVE 状态。开发者必须重写该方法,并在此通过 setMainRoute()方法配置默认展示的 AbilitySlice。
- onActive():类似于 Android 中的 onResume(),Page Ability 会在进入 INACTIVE 状态后来到前台,然后系统调用此回调。Page Ability 在此之后进入 ACTIVE 状态,该状态是应用与用户交互的状态。Page Ability 将保持在此状态,除非某类事件发生导致 Page Ability 失去焦点,比如用户单击返回键或导航到其他 Page Ability。当此类事件发生时,会触发 Page Ability 回到 INACTIVE 状态,系统将调用 onInactive()回调。此后,Page Ability 可能重新回到 ACTIVE 状态,系统将再次调用 onActive()回调。因此,开发者通常需要成对实现 onActive()和 onInactive(),并在 onActive()中获取在 onInactive()中被释放的资源。

- onInactive()：当 Page Ability 失去焦点时，系统将调用此回调，此后 Page Ability 进入 INACTIVE 状态，这类似于 Android 中的 onPause() 和 onStop() 的集合体。开发者可以在此回调中实现 Page Ability 失去焦点时所需要实现的操作。
- onBackground()：当 Page Ability 不再对用户可见时，系统将调用此回调通知开发者进行相应的资源释放，此后该 Page Ability 将进入 BACKGROUND 状态。开发者应该在此回调中释放 Page Ability 不可见时无用的资源，也可在此回调中执行较为耗时的状态保存操作。
- onForeground()：处于 BACKGROUND 状态的 Page Ability 仍然驻留在内存中，当重新回到前台时（比如用户重新导航到此 Page Ability），系统将先调用 onForeground() 回调通知开发者，然后 Page Ability 的生命周期状态将会回到 INACTIVE 状态。开发者应当在此回调中重新申请在 onBackground() 中释放的资源，最后 Page Ability 的生命周期状态进一步回到 ACTIVE 状态，系统将通过 onActive() 回调通知开发者。这和 Android 中的 onRestart() 方法类似。
- onStop()：系统将要销毁 Page Ability 时，将会触发此回调函数，通知开发者释放系统资源。这和 Android 中的 onDestroy() 方法类似。

4.2.5 AbilitySlice 生命周期

Page Ability 有自己的生命周期，作为 Page Ability 组成单元的 AbilitySlice 也有自己的生命周期，其生命周期是依托于其所属 Page Ability 生命周期的。AbilitySlice 和 Page Ability 具有相同的生命周期状态和同名的回调，当 Page Ability 生命周期发生变化时，它的 AbilitySlice 也会发生相同的生命周期变化。此外，当同一 Page Ability 中的不同 AbilitySlice 之间导航时，AbilitySlice 还具有独立于 Page Ability 的生命周期变化，此时 Page Ability 的生命周期状态并不会改变。

具体来说，当 AbilitySlice 处于前台且具有焦点时，其生命周期状态随着所属 Page Ability 的生命周期状态的变化而变化。当一个 Page Ability 拥有多个 AbilitySlice 时，例如：MyAbility 下有 FirstAbilitySlice 和 SecondAbilitySlice 两个 AbilitySlice，当 FirstAbilitySlice 处于前台且获得焦点并即将导航到 SecondAbilitySlice 时，其生命周期状态的变化顺序为：

（1）FirstAbilitySlice 从 ACTIVE 状态变为 INACTIVE 状态。

（2）SecondAbilitySlice 则首先从 INITIAL 状态变为 INACTIVE 状态，然后变为 ACTIVE 状态。

（3）FirstAbilitySlice 从 INACTIVE 状态变为 BACKGROUND 状态。

对应两个 AbilitySlice 的生命周期方法回调顺序为：

```
FirstAbilitySlice.onInactive() --> SecondAbilitySlice.onStart() -->
SecondAbilitySlice.onActive() --> FirstAbilitySlice.onBackground()
```

在整个生命周期中，MyAbility 始终处于 ACTIVE 状态。但是，当 Page Ability 被系统销毁时，不单是当前处于前台的 AbilitySlice，其所有已实例化的 AbilitySlice 都将跟着一起销毁。

由于 AbilitySlice 承载具体的页面，因此开发者必须重写 AbilitySlice 的 onStart() 回调，并在此方法中通过 setUIContent() 方法设置页面，setUIContent() 需要传入一个 ComponentContainer，通常为一个 XML 布局文件，该布局文件通过项目自身的 ResourceTable 进行引用，详细代码如程序清单 4-4 所示。

程序清单 4-4：hmos\ch04\01\AbilityTest\entry\src\main\java\com\example\abilitytest\slice\MainAbilitySlice.java

```
1  @Override
2  protected void onStart(Intent intent) {
3      super.onStart(intent);
4      setUIContent(ResourceTable.Layout_ability_main);
5  }
```

需要注意的是，在引用 ResourceTable 时，要区分 ohos.global.systemres.ResourceTable 与 com.xxx.xxx.ResourceTable，前者是 HarmonyOS 系统自身的 ResourceTable，后者才是此处需要引用的项目自身的 ResourceTable，如图 4-10 所示，我们需要引用的为第一个 ResourceTable。

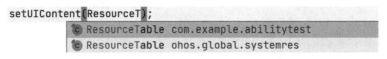

图 4-10　导入 ResourceTable

在引用 xml 布局文件时，被引用的文件的名称由文件所在的文件夹名加上文件名组成，例如上面示例代码中的 Layout_ability_main 表示的就是 resources 文件夹下的 base 文件夹下的 layout 文件夹里面名为 ability_main.xml 的资源文件，如图 4-11 所示。

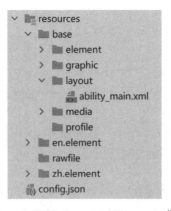

图 4-11　ResourceTable.Layout_ability_main 指向的文件

4.2.6 AbilitySlice 间的导航

AbilitySlice 间的导航分为同一 Page Ability 内的导航和不同 Page Ability 内的导航。同一 Page Ability 内的导航可以通过 present() 和 presentForResult() 方法实现。present() 实现的是无返回结果的导航,而 presentForResult() 实现的是有返回结果的导航。用 present() 实现导航的操作代码较为简单,只使用 present() 方法即可,详细代码如程序清单 4-5 所示。

程序清单 4-5：hmos\ch04\01\AbilityTest\entry\src\main\java\com\example\abilitytest\slice\MainAbilitySlice.java

```
present(new TargetSlice(), new Intent());
```

当使用 presentForResult() 实现导航时,用户从导航目标 AbilitySlice 返回时,系统将回调 onResult() 来接收和处理返回结果,开发者需要重写该方法。返回结果由导航目标 AbilitySlice 在其生命周期内通过 setResult() 进行设置。原 AbilitySlice 里实现导航的示例代码如程序清单 4-6 所示。

程序清单 4-6：hmos\ch04\01\AbilityTest\entry\src\main\java\com\example\abilitytest\slice\MainAbilitySlice.java

```
presentForResult(new TargetSlice(), new Intent(), 0);
                                    //第 3 个参数为请求码,可自由设定
```

我们还要重写 onResult() 方法来接收和处理返回结果,其中 requestCode 的值与启动导航的 presentForResult() 中的第 3 个参数一致,用来识别是由哪个 presentForResult() 实现导航的,resultIntent 的值为导航目标 AbilitySlice 里设置的用于返回的 Intent,可以通过 resultIntent.getXXXParam() 的方式获取导航目标 AbilitySlice 里设置的返回值。示例代码如程序清单 4-7 所示。

程序清单 4-7：hmos\ch04\01\AbilityTest\entry\src\main\java\com\example\abilitytest\slice\MainAbilitySlice.java

```
1  @Override
2  protected void onResult(int requestCode, Intent resultIntent) {
3      //处理返回结果
4      if (requestCode == 0) {
5          ....
6      }
7  }
```

在导航的目标 AbilitySlice 里,需通过 setResult() 方法设置返回结果,示例代码如程序清单 4-8 所示。

程序清单 4-8：hmos\ch04\01\AbilityTest\entry\src\main\java\
com\example\abilitytest\slice\MainAbilitySlice2.java

```
1    Intent resultIntent = new Intent();
2    resultIntent.setParam("A", 1);
                        //自定义的返回值,第一个参数为 key 值,第二个参数为 value
3    setResult(resultIntent);
```

4.2.7 不同 Page Ability 间的导航

AbilitySlice 作为 Page Ability 的内部单元,以 Action 的形式对外暴露,因此可以通过配置 Intent 的 Action 导航到目标 AbilitySlice。Page Ability 间的导航可以使用 startAbility()或 startAbilityForResult()方法,获得返回结果的回调为 onAbilityResult()。在 Ability 中调用 setResult()可以设置返回结果。值得注意的是,在 Ability 中调用的 setResult()方法和在 AbilitySlice 中调用的 setResult()方法是不一样的,Ability 中调用的 setResult()方法比 AbilitySlice 中调用的 setResult()方法需多传递一个 resultCode 参数,该参数由开发者自定义,可用来识别当前动作是从哪个 Ability 返回的。

当 Intent 用于发起请求时,根据指定元素的不同,分为两种类型:一种是当同时指定了 BundleName 与 AbilityName 时,则根据 Ability 的全称(例如"com.example.abilitytest.TestAbility")直接启动应用;另一种是未同时指定 BundleName 和 AbilityName,则根据 Operation 中的其他属性启动应用。

1. 根据 Ability 的全称启动

通过构造包含 BundleName 与 AbilityName 的 Operation 对象,可以启动一个 Ability 并导航到该 Ability。导航的目标 Ability 可以在其 onStart()回调的参数中获得请求方传递的 Intent 对象。示例代码如程序清单 4-9 所示。

程序清单 4-9：hmos\ch04\01\AbilityTest\entry\src\main\java\
com\example\abilitytest\slice\MainAbilitySlice.java

```
1    Intent intent = new Intent();
2    //通过 Intent 中的 OperationBuilder 类构造 operation 对象
3    Operation operation = new Intent.OperationBuilder()
4            .withDeviceId("")
5            .withBundleName("com.example.abilitytest")
6            .withAbilityName("TestAbility")
7            .build();
8    //把 operation 设置到 intent 中
9    intent.setOperation(operation);
10   startAbility(intent);
```

其中,withDeviceId()方法设置的是指定设备标识,设置为空串表示当前设备。

withBundleName()设置的是应用包名,withAbilityName()设置的是需要启动的 Ability 名称。

2. 根据 Operation 的其他属性启动

在某些情景下,开发者需要在应用中使用其他应用提供的某种功能,而不知道提供该功能的具体是哪一个应用。例如,开发者需要通过浏览器打开一个链接,而不关心用户最终选择哪一个浏览器应用,则可以通过 Operation 的除 BundleName 与 AbilityName 之外的其他属性描述需要的能力。如果设备上存在多个应用提供同种能力,系统则弹出候选列表,由用户选择由哪个应用处理请求。

- 请求方:

对于请求方,在 Ability 中构造 Intent 以及包含 Action 的 Operation 对象,并调用 startAbility()方法发起请求。示例代码如程序清单 4-10 所示。

程序清单 4-10: hmos\ch04\01\AbilityTest\entry\src\main\java\com\example\abilitytest\slice\MainAbilitySlice.java

```
1  Intent intent = new Intent();
2  Operation operation = new Intent.OperationBuilder()
3          .withAction("action.mytest")
4          .build();
5  intent.setOperation(operation);
6  startAbility(intent);
```

- 处理方:

对于处理请求的对象,首先需要在配置文件(config.json)中声明对外提供的能力,以便系统据此找到自身并作为候选的请求处理者。示例代码如程序清单 4-11 所示。

程序清单 4-11: hmos\ch04\01\AbilityTest\entry\src\main\config.json

```
1   {
2       "module": {
3           ...
4           "abilities": [
5               {
6                   ...
7                   "skills":[
8                       {
9                           "actions":[
10                              ...
11                              "action.mytest"
12                          ]
13                      }
```

```
14            ]
15            ...
16          }
17        ]
18        ...
19      }
20      ...
21 }
```

在 Ability 的 onStart()方法中通过使用 addActionRoute()方法配置路由以便支持以此 action 导航到对应的 AbilitySlice。示例代码如程序清单 4-12 所示。

程序清单 4-12：hmos\ch04\01\AbilityTest\entry\src\main\java\
com\example\abilitytest\MainAbility.java

```
1  @Override
2  protected void onStart(Intent intent) {
3      ...
4      addActionRoute("action.mytest", SecondSlice.class.getName());
5      ...
6  }
```

当开发者需要实现对动作的返回进行处理时，可以使用 startAbilityForResult()方法导航，并实现和重写相应方法。这里以根据 Ability 的全称启动应用的方法为例实现处理。

首先在请求方先通过 startAbilityForResult()启动导航，相比 startAbility()需多传递一个 resultCode 参数，示例代码如程序清单 4-13 所示。

程序清单 4-13：hmos\ch04\01\AbilityTest\entry\src\main\java\com\
example\abilitytest\slice\MainAbilitySlice.java

```
1  private void private void startSecondAbility() {
2      Intent intent = new Intent();
3      Operation operation = new Intent.OperationBuilder()
4          .withDeviceId("")
5          .withBundleName("com.example.abilitytest")
6          .withAbilityName("TestAbility")
7          .build();
8      intent.setOperation(operation);
9      startAbilityForResult(intent, 0);
10  }
11
```

```
12   @Override
13   protected void onAbilityResult(int requestCode, int resultCode, Intent
     resultData) {
14       switch (requestCode) {
15           case 0:
16               //对数据进行处理
17               ...
18               return;
19           default:
20               ...
21       }
22   }
```

在导航目标 Ability 中处理请求,并调用 setResult()方法暂存返回结果,示例代码如程序清单 4-14 所示。

程序清单 4-14: hmos\ch04\01\AbilityTest\entry\src\main\java\com\example\abilitytest\slice\MainAbilitySlice2.java

```
1   @Override
2   protected void onActive() {
3       ...
4       Intent resultIntent = new Intent();
5       setResult(0, resultIntent);//0 为当前 Ability 销毁后返回的 resultCode
6       ...
7   }
```

4.2.8 用户注册案例

下面用一个完整的注册案例讲解 Page Ability 间的数据传输。程序主界面如图 4-12(a)所示,当单击所在地时,程序会弹出一个 CommonDialog 对话框以供选择地区(图 4-12(b)),当单击某一地区后,会弹出一个 ToastDialog 对话框,用来提示你选择了哪一个地区(图 4-12(c))。填写完相应的信息后,单击"注册"按钮,系统会对用户填写的信息进行验证,如用户名未填写,则会弹出 ToastDialog 对话框(图 4-12(d));如果两次密码不一致,则弹出 ToastDialog 对话框(图 4-12(e))。如果用户的注册信息符合要求,则跳转到注册成功界面(图 4-12(f)所示)。

项目中涉及多个布局文件和界面,它们各自的作用和关系如图 4-13 所示。

下面详细介绍这个程序的开发过程。首先是注册界面的设计,详细代码如程序清单 4-15 所示。

(a) 运行界面　　　　　(b) 选择所在地　　　　　(c) 选择所在地提示

(d) 账号验证提示　　　　(e) 密码验证提示　　　　(f) 注册成功跳转界面

图 4-12　用户注册案例

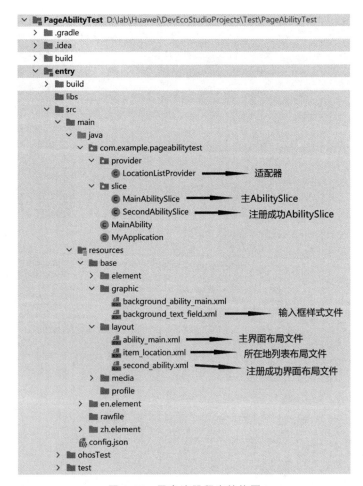

图 4-13 用户注册程序结构图

程序清单 4-15: hmos\ch04\02\PageAbilityTest\entry\src\main\resources\base\layout\ability_main.xml

```
1   <?xml version="1.0" encoding="utf-8"?>
2   <DirectionalLayout
3       xmlns:ohos="http://schemas.huawei.com/res/ohos"
4       ohos:height="match_parent"
5       ohos:width="match_parent"
6       ohos:alignment="top|center"
7       ohos:orientation="vertical"
8       ohos:padding="20vp">
9
10      <Text
11          ohos:height="match_content"
12          ohos:width="match_content"
```

```
13          ohos:background_element="$graphic:background_ability_main"
14          ohos:layout_alignment="horizontal_center"
15          ohos:text="欢迎注册倚动资源网"
16          ohos:text_size="25vp"/>
17
18      <Component
19          ohos:height="1vp"
20          ohos:width="match_parent"
21          ohos:background_element="#000000"
22          ohos:margin="5vp"
23          ohos:top_margin="9vp"/>
24
25      <DirectionalLayout
26          ohos:height="match_content"
27          ohos:width="match_parent"
28          ohos:orientation="horizontal"
29          ohos:top_margin="9vp"
30          ohos:top_padding="9vp">
31
32          <Text
33              ohos:height="match_content"
34              ohos:width="match_parent"
35              ohos:text="账        号:"
36              ohos:text_alignment="center"
37              ohos:text_size="18vp"
38              ohos:weight="1"/>
39
40          <TextField
41              ohos:id="$+id:textfield_id"
42              ohos:height="match_content"
43              ohos:width="match_parent"
44              ohos:background_element="$graphic:background_text_field"
45              ohos:hint="输入你的账号"
46              ohos:left_padding="7vp"
47              ohos:right_margin="15vp"
48              ohos:text_alignment="vertical_center"
49              ohos:text_size="18vp"
50              ohos:weight="2"/>
51
52      </DirectionalLayout>
53
54      <DirectionalLayout
55          ohos:height="match_content"
```

```
56              ohos:width="match_parent"
57              ohos:orientation="horizontal"
58              ohos:top_margin="9vp"
59              ohos:top_padding="9vp">
60
61          <Text
62              ohos:height="match_content"
63              ohos:width="match_parent"
64              ohos:text="密        码  :"
65              ohos:text_alignment="center"
66              ohos:text_size="18vp"
67              ohos:weight="1"/>
68
69          <TextField
70              ohos:id="$+id:textfield_password_1"
71              ohos:height="match_content"
72              ohos:width="match_parent"
73              ohos:background_element="$graphic:background_text_field"
74              ohos:hint="输入你的密码"
75              ohos:left_padding="7vp"
76              ohos:right_margin="15vp"
77              ohos:text_alignment="vertical_center"
78              ohos:text_input_type="pattern_password"
79              ohos:text_size="18vp"
80              ohos:weight="2"/>
81
82      </DirectionalLayout>
83
84      <DirectionalLayout
85          ohos:height="match_content"
86          ohos:width="match_parent"
87          ohos:orientation="horizontal"
88          ohos:top_margin="9vp"
89          ohos:top_padding="9vp">
90
91          <Text
92              ohos:height="match_content"
93              ohos:width="match_parent"
94              ohos:text="确认密码  :"
95              ohos:text_alignment="center"
96              ohos:text_size="18vp"
97              ohos:weight="1"/>
98
```

```xml
        <TextField
            ohos:id="$+id:textfield_password_2"
            ohos:height="match_content"
            ohos:width="match_parent"
            ohos:background_element="$graphic:background_text_field"
            ohos:hint="两次密码必须一致"
            ohos:left_padding="7vp"
            ohos:right_margin="15vp"
            ohos:text_alignment="vertical_center"
            ohos:text_input_type="pattern_password"
            ohos:text_size="18vp"
            ohos:weight="2"/>

    </DirectionalLayout>

    <DirectionalLayout
        ohos:height="match_content"
        ohos:width="match_parent"
        ohos:orientation="horizontal"
        ohos:top_margin="9vp"
        ohos:top_padding="9vp">

        <Text
            ohos:height="match_content"
            ohos:width="match_parent"
            ohos:text="性        别："
            ohos:text_alignment="center"
            ohos:text_size="18vp"
            ohos:weight="1"/>

        <RadioContainer
            ohos:id="$+id:radio_container"
            ohos:height="match_content"
            ohos:width="match_parent"
            ohos:orientation="horizontal"
            ohos:right_margin="15vp"
            ohos:weight="2">

            <RadioButton
                ohos:height="match_content"
                ohos:width="match_content"
                ohos:left_padding="25vp"
                ohos:text="男"
```

```
142                ohos:text_size="18vp"/>
143
144            <RadioButton
145                ohos:height="match_content"
146                ohos:width="match_content"
147                ohos:left_padding="25vp"
148                ohos:text="女"
149                ohos:text_size="18vp"/>
150        </RadioContainer>
151
152    </DirectionalLayout>
153
154    <DirectionalLayout
155        ohos:height="match_content"
156        ohos:width="match_parent"
157        ohos:orientation="horizontal"
158        ohos:top_margin="9vp"
159        ohos:top_padding="9vp">
160
161        <Text
162            ohos:height="match_content"
163            ohos:width="match_parent"
164            ohos:text="所 在 地 :"
165            ohos:text_alignment="center"
166            ohos:text_size="18vp"
167            ohos:weight="1"/>
168
169        <Button
170            ohos:id="$+id:button_location"
171            ohos:height="match_content"
172            ohos:width="match_parent"
173            ohos:background_element="$graphic:background_text_field"
174            ohos:hint="请选择所在地"
175            ohos:left_padding="7vp"
176            ohos:right_margin="15vp"
177            ohos:text_alignment="vertical_center"
178            ohos:text_size="18vp"
179            ohos:weight="2"/>
180
181    </DirectionalLayout>
182
183    <Button
184        ohos:id="$+id:button_submit"
```

```
185            ohos:height="match_content"
186            ohos:width="match_parent"
187            ohos:background_element="#00ff00"
188            ohos:margin="25vp"
189            ohos:padding="9vp"
190            ohos:text="注     册"
191            ohos:text_alignment="horizontal_center"
192            ohos:text_size="18vp"/>
193
194   </DirectionalLayout>
```

其中 RadioContainer 表示单选按钮组,RadioButton 表示单选按钮,放在同一个组里面的单选按钮方能互斥,每次只能选其中的一个。默认情况下,单选按钮是垂直摆放的,如果需要水平摆放,可以通过设置方向为水平,即 ohos:orientation="horizontal"。

界面设计好以后,需要对界面中的 3 个按钮添加相应的时间处理,首先为单选按钮组添加事件处理,详细代码如程序清单 4-16 所示。

程序清单 4-16:hmos\ch04\02\PageAbilityTest\entry\src\main\java\com\example\pageabilitytest\slice\MainAbilitySlice.java

```
1    private void setListener() {
2        radioContainer.setMarkChangedListener(new RadioContainer.
     CheckedStateChangedListener() {
3            @Override
4            public void onCheckedChanged(RadioContainer radioContainer, int i) {
5                if (i == 0) {
6                    sex = "男";
7                } else {
8                    sex = "女";
9                }
10           }
11       });
12   }
```

接下来为所在地按钮添加事件处理,详细代码如程序清单 4-17 所示。

程序清单 4-17:hmos\ch04\02\PageAbilityTest\entry\src\main\java\com\example\pageabilitytest\slice\MainAbilitySlice.java

```
1    private void setListener() {
2        ...
3        button_location.setClickedListener(new Component.ClickedListener() {
4            @Override
5            public void onClick(Component component) {
```

```
 6              CommonDialog dialog = new CommonDialog(getContext());
 7              dialog.setTitleText("   选择所在地");
 8              dialog.setCornerRadius(30);
 9              dialog.setSize(750, ComponentContainer.LayoutConfig.MATCH_
    CONTENT);
10              ListContainer listContainer = new ListContainer(getContext());
11              listContainer.setWidth(ComponentContainer.LayoutConfig.
    MATCH_PARENT);
12              listContainer.setHeight(ComponentContainer.LayoutConfig.
    MATCH_CONTENT);
13              listContainer.setPadding(10, 10, 10, 50);
14              LocationListProvider provider=new LocationListProvider
    (provinces, MainAbilitySlice.this);
15              listContainer.setItemProvider(provider);
16              listContainer.setItemClickedListener(new ListContainer.
    ItemClickedListener() {
17                  @Override
18                  public void onItemClicked(ListContainer listContainer,
    Component component, int i, long l) {
19                      new ToastDialog(getContext())
20                              .setText("你选择了" + provinces.get(i))
21                              .show();
22                      province = provinces.get(i);
23                  }
24              });
25              dialog.setContentCustomComponent(listContainer);
26              dialog.setButton(IDialog.BUTTON1, "取消", new IDialog.
    ClickedListener() {
27                  @Override
28                  public void onClick(IDialog iDialog, int i) {
29                      iDialog.destroy();
30                  }
31              });
32              dialog.setButton(IDialog.BUTTON2, "确定", new IDialog.
    ClickedListener() {
33                  @Override
34                  public void onClick(IDialog iDialog, int i) {
35                      if ("".equals(province)) {
36                          new ToastDialog(getContext())
37                                  .setText("你还没选呢")
38                                  .show();
39                      } else {
40                          button_location.setText(province);
```

```
41                iDialog.destroy();
42            }
43        }
44    });
45    dialog.show();
46  }
47 });
48 }
```

在所在地按钮的事件处理中使用了自定义的适配器 LocationListProvider，详细代码如程序清单 4-18 所示。

程序清单 4-18：hmos\ch04\02\PageAbilityTest\entry\src\main\java\com\example\pageabilitytest\provider\LocationListProvider.java

```
1  public class LocationListProvider extends BaseItemProvider {
2      private List<String> provinces = new ArrayList<>();
3      private AbilitySlice slice;
4
5      public LocationListProvider(List<String> provinces, AbilitySlice slice) {
6          this.provinces = provinces;
7          this.slice = slice;
8      }
9
10     @Override
11     public int getCount() {
12         return provinces == null ? 0 : provinces.size();
13     }
14
15     @Override
16     public Object getItem(int i) {
17         return provinces.get(i);
18     }
19
20     @Override
21     public long getItemId(int i) {
22         return i;
23     }
24
25     @Override
26     public Component getComponent(int i, Component component, ComponentContainer componentContainer) {
27         final Component cpt;
```

```
28          if (component == null) {
29              cpt = LayoutScatter.getInstance(slice).parse(ResourceTable.
    Layout_item_location, null, false);
30          } else {
31              cpt = component;
32          }
33          Text text = (Text) cpt.findComponentById(ResourceTable.Id_item_
    index);
34          text.setText(provinces.get(i));
35          return cpt;
36      }
37  }
```

最后为注册按钮添加事件处理,详细代码如程序清单4-19所示。

```
1   private void setListener() {
2       ...
3       button_submit.setClickedListener(new Component.ClickedListener() {
4           @Override
5           public void onClick(Component component) {
6               string_id = textfield_id.getText();
7               string_password_1 = textfield_password_1.getText();
8               string_password_2 = textfield_password_2.getText();
9               if (!"".equals(string_id)) {
10                  if (string_password_1.equals(string_password_2)) {
11                      System.out.println("success");
12                      Intent intent = new Intent();
13                      intent.setParam("id", string_id);
14                      intent.setParam("sex", sex);
15                      intent.setParam("province", province);
16                      present(new SecondAbilitySlice(), intent);
17                  } else {
18                      new ToastDialog(getContext())
19                              .setText("两次密码不一致")
20                              .show();
21                  }
22              } else {
23                  new ToastDialog(getContext())
24                          .setText("请输入账号")
25                          .show();
26              }
```

```
27          }
28      });
29 }
```

在上面的注册时间处理中注册成功后会跳转到注册成功界面，即 SecondAbilitySlice 所呈现的界面，其布局文件详细代码如程序清单 4-20 所示。

程序清单 4-20：hmos\ch04\02\PageAbilityTest\entry\src\
main\resources\base\layout\second_ability.xml

```xml
1  <?xml version="1.0" encoding="utf-8"?>
2  <DirectionalLayout
3      xmlns:ohos="http://schemas.huawei.com/res/ohos"
4      ohos:height="match_parent"
5      ohos:width="match_parent"
6      ohos:orientation="vertical">
7
8      <Text
9          ohos:height="match_content"
10         ohos:width="match_content"
11         ohos:layout_alignment="horizontal_center"
12         ohos:padding="20vp"
13         ohos:text="恭喜您,注册成功!"
14         ohos:text_size="25vp"/>
15
16     <DirectionalLayout
17         ohos:height="match_content"
18         ohos:width="match_parent"
19         ohos:orientation="horizontal"
20         ohos:padding="15vp"
21         ohos:top_margin="9vp">
22
23         <Text
24             ohos:height="match_content"
25             ohos:width="match_content"
26             ohos:text="您的账号为:"
27             ohos:text_size="18vp"/>
28
29         <Text
30             ohos:id="$+id:text_id"
31             ohos:height="match_content"
32             ohos:width="match_content"
33             ohos:text_size="18vp"/>
34     </DirectionalLayout>
```

```xml
35
36      <DirectionalLayout
37          ohos:height="match_content"
38          ohos:width="match_parent"
39          ohos:orientation="horizontal"
40          ohos:padding="15vp"
41          ohos:top_margin="9vp">
42
43          <Text
44              ohos:height="match_content"
45              ohos:width="match_content"
46              ohos:text="您的性别为:"
47              ohos:text_size="18vp"/>
48
49          <Text
50              ohos:id="$+id:text_sex"
51              ohos:height="match_content"
52              ohos:width="match_content"
53              ohos:text_size="18vp"/>
54      </DirectionalLayout>
55
56      <DirectionalLayout
57          ohos:height="match_content"
58          ohos:width="match_parent"
59          ohos:orientation="horizontal"
60          ohos:padding="15vp"
61          ohos:top_margin="9vp">
62
63          <Text
64              ohos:height="match_content"
65              ohos:width="match_content"
66              ohos:text="所在城市为:"
67              ohos:text_size="18vp"/>
68
69          <Text
70              ohos:id="$+id:text_location"
71              ohos:height="match_content"
72              ohos:width="match_content"
73              ohos:text_size="18vp"/>
74      </DirectionalLayout>
75 </DirectionalLayout>
```

SecondAbilitySlice 的详细代码如程序清单 4-21 所示。

程序清单 4-21：hmos\ch04\02\PageAbilityTest\entry\src\main\java\com\example\pageabilitytest\slice\SecondAbilitySlice.java

```java
public class SecondAbilitySlice extends AbilitySlice {
    private String province = "", sex = "", string_id = "";
    private Text text_id, text_sex, text_location;

    @Override
    protected void onStart(Intent intent) {
        super.onStart(intent);
        setUIContent(ResourceTable.Layout_second_ability);
        string_id = intent.getStringParam("id");
        sex = intent.getStringParam("sex");
        province = intent.getStringParam("province");
        initLayout();
        text_id.setText(string_id);
        text_sex.setText(sex);
        text_location.setText(province);
    }

    private void initLayout() {
        text_id = (Text) findComponentById(ResourceTable.Id_text_id);
        text_sex = (Text) findComponentById(ResourceTable.Id_text_sex);
        text_location = (Text) findComponentById(ResourceTable.Id_text_location);
    }
}
```

MainAbilitySlice 的完整代码如程序清单 4-22 所示。

程序清单 4-22：hmos\ch04\02\PageAbilityTest\entry\src\main\java\com\example\pageabilitytest\slice\MainAbilitySlice.java

```java
public class MainAbilitySlice extends AbilitySlice {
    private Button button_location, button_submit;
    private List<String> provinces = new ArrayList<>(Arrays.asList("南昌", "赣州", "九江"));
    private String province = "", sex = "", string_id = "", string_password_1 = "", string_password_2 = "";
    private TextField textfield_id, textfield_password_1, textfield_password_2;
    private RadioContainer radioContainer;

    @Override
```

```
9    public void onStart(Intent intent) {
10       super.onStart(intent);
11       super.setUIContent(ResourceTable.Layout_ability_main);
12       initLayout();
13       setListener();
14   }
15
16   private void setListener() {
17       button_location.setClickedListener(new Component.ClickedListener() {
18           @Override
19           public void onClick(Component component) {
20               CommonDialog dialog = new CommonDialog(getContext());
21               dialog.setTitleText("   选择所在地");
22               dialog.setCornerRadius(30);
23               dialog.setSize(750, ComponentContainer.LayoutConfig.MATCH_CONTENT);
24               ListContainer listContainer = new ListContainer(getContext());
25               listContainer.setWidth(ComponentContainer.LayoutConfig.MATCH_PARENT);
26               listContainer.setHeight(ComponentContainer.LayoutConfig.MATCH_CONTENT);
27               listContainer.setPadding(10, 10, 10, 50);
28               LocationListProvider provider = new LocationListProvider(provinces, MainAbilitySlice.this);
29               listContainer.setItemProvider(provider);
30               listContainer.setItemClickedListener(new ListContainer.ItemClickedListener() {
31                   @Override
32                   public void onItemClicked(ListContainer listContainer, Component component, int i, long l) {
33                       new ToastDialog(getContext())
34                               .setText("你选择了" + provinces.get(i))
35                               .show();
36                       province = provinces.get(i);
37                   }
38               });
39               dialog.setContentCustomComponent(listContainer);
40               dialog.setButton(IDialog.BUTTON1, "取消", new IDialog.ClickedListener() {
41                   @Override
42                   public void onClick(IDialog iDialog, int i) {
43                       iDialog.destroy();
```

```
44                    }
45                });
46                dialog.setButton(IDialog.BUTTON2, "确定", new IDialog.
   ClickedListener() {
47                    @Override
48                    public void onClick(IDialog iDialog, int i) {
49                        if ("".equals(province)) {
50                            new ToastDialog(getContext())
51                                    .setText("你还没选呢")
52                                    .show();
53                        } else {
54                            button_location.setText(province);
55                            iDialog.destroy();
56                        }
57
58                    }
59                });
60                dialog.show();
61            }
62        });
63        radioContainer.setMarkChangedListener(new RadioContainer.
   CheckedStateChangedListener() {
64            @Override
65            public void onCheckedChanged(RadioContainer radioContainer,
   int i) {
66                if (i == 0) {
67                    sex = "男";
68                } else {
69                    sex = "女";
70                }
71            }
72        });
73        button_submit.setClickedListener(new Component.ClickedListener() {
74            @Override
75            public void onClick(Component component) {
76                string_id = textfield_id.getText();
77                string_password_1 = textfield_password_1.getText();
78                string_password_2 = textfield_password_2.getText();
79                if (!"".equals(string_id)) {
80                    if (string_password_1.equals(string_password_2)) {
81                        System.out.println("success");
82                        Intent intent = new Intent();
83                        intent.setParam("id", string_id);
```

```
84                        intent.setParam("sex", sex);
85                        intent.setParam("province", province);
86                        present(new SecondAbilitySlice(), intent);
87
88                   } else {
89                       new ToastDialog(getContext())
90                               .setText("两次密码不一致")
91                               .show();
92                   }
93               } else {
94                   new ToastDialog(getContext())
95                           .setText("请输入账号")
96                           .show();
97               }
98
99           }
100      });
101    }
102
103    private void initLayout() {
104        button_location = (Button) findComponentById(ResourceTable.Id_button_location);
105        textfield_id = (TextField) findComponentById(ResourceTable.Id_textfield_id);
106        textfield_password_1 = (TextField) findComponentById(ResourceTable.Id_textfield_password_1);
107        textfield_password_2 = (TextField) findComponentById(ResourceTable.Id_textfield_password_2);
108        radioContainer = (RadioContainer) findComponentById(ResourceTable.Id_radio_container);
109        button_submit = (Button) findComponentById(ResourceTable.Id_button_submit);
110    }
111
112    @Override
113    public void onActive() {
114        super.onActive();
115    }
116
117    @Override
118    public void onForeground(Intent intent) {
119        super.onForeground(intent);
120    }
121 }
```

4.3 Service Ability

之前章节中涉及的所有工作都是通过 Page Ability 完成的,根据 Page Ability 的生命周期,程序中每次只有一个 Page Ability 处于激活状态,并且 Page Ability 的执行时间有限,不能做一些比较耗时的操作。当需要多种工作同时进行时,如一边听音乐,一边浏览网页,则比较困难。针对这种情况,HarmonyOS 为我们提供了 Service Ability。

Service Ability 与 Android 中的 Service 类似。Service Ability 不提供用户交互界面,而是一直在 HarmonyOS 的后台运行,相当于一个没有图形界面的 Page Ability 程序,它不能自己直接运行,需要由其他应用或 Ability 启动。

4.3.1 创建 Service Ability

下面创建一个 Service Ability,在当前工程的主目录(entry>src>main>java>com.xxx.xxx)上右击,从弹出的快捷菜单中选择 New>Ability>Empty Service Ability,如图 4-14 所示。

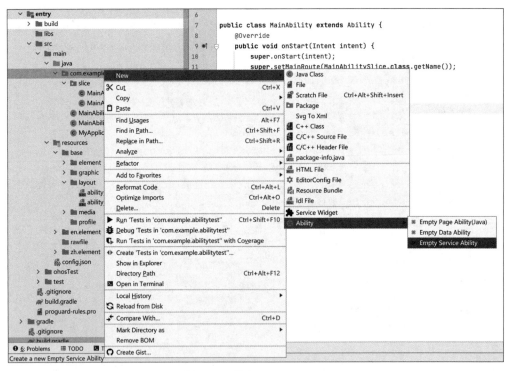

图 4-14 创建 Service Ability

然后,进入如图 4-15 所示的界面。

其中,Service Name 框里填写需要自定义的 Service Ability 的名称,Package name 框里填写新建的 Service Ability 需要放在哪个包下,如果填写的包名不存在,HUAWEI

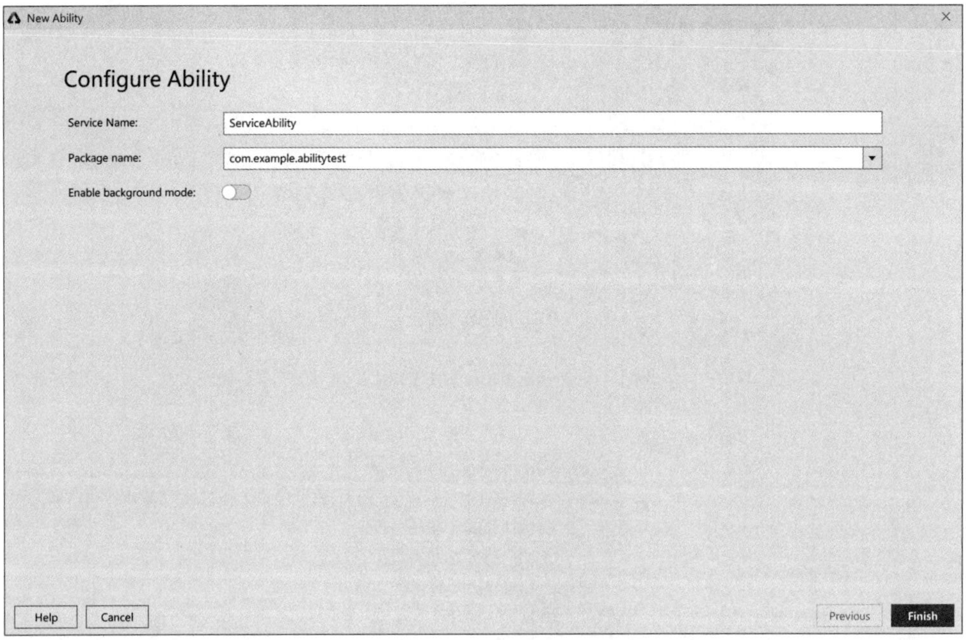

图 4-15 新建 Service Ability 相关信息

DevEco Studio 会自动创建该包。而 Enable background mode 表示后台模式，如果开发者打开这个开关（图 4-16），就表示新建的 Service 要在后台运行，开发者还可以自己选择该 Service 要在后台进行什么类型的操作。

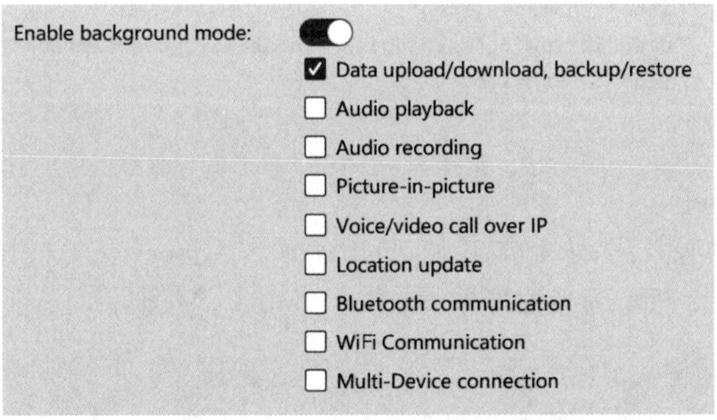

图 4-16 Enable background mode 中的相关选项信息

图 4-16 翻译成中文如图 4-17 所示。

下面直接创建。不打开 Enable background mode 选项，单击 Finish 按钮。创建 Service Ability 不会生成 AbilitySlice，如图 4-18 所示。

在创建完 Service Ability 的同时，HUAWEI DevEco Studio 也会在配置文件（config.json）中注册该 Service Ability，相比于 Page Ability，Service Ability 的属性要少一些，而且 type 的属性值是 service。具体代码如图 4-19 所示。

Service Ability 也是一种 Ability，Ability 为 Service Ability 提供了以下生命周期方

图 4-17 Enable background mode 中的相关选项信息（中文）

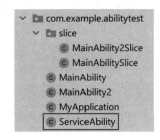

图 4-18 新建 Service Ability 的相关文件

```
{
  "name": "com.example.abilitytest.ServiceAbility",
  "icon": "$media:icon",
  "description": "$string:serviceability_description",
  "type": "service"
}
```

图 4-19 新建 Service Ability 的配置信息

法，用户可以重写这些方法，来添加其他 Ability 请求与 Service Ability 交互时的处理方法。如程序清单 4-23 所示为新建好的 Service Ability 的代码。

程序清单 4-23：hmos\ch04\01\AbilityTest\entry\src\main\java\com\example\abilitytest\ServiceAbility.java

```
1  package com.example.abilitytest;
2
3  import ohos.aafwk.ability.Ability;
4  import ohos.aafwk.content.Intent;
5  import ohos.rpc.IRemoteObject;
6  import ohos.hiviewdfx.HiLog;
7  import ohos.hiviewdfx.HiLogLabel;
8
```

```
9   public class ServiceAbility extends Ability {
10      private static final HiLogLabel LABEL_LOG = new HiLogLabel(3,
    0xD001100, "Demo");
11
12      @Override
13      public void onStart(Intent intent) {
14          HiLog.error(LABEL_LOG, "ServiceAbility::onStart");
15          super.onStart(intent);
16      }
17
18      @Override
19      public void onBackground() {
20          super.onBackground();
21          HiLog.info(LABEL_LOG, "ServiceAbility::onBackground");
22      }
23
24      @Override
25      public void onStop() {
26          super.onStop();
27          HiLog.info(LABEL_LOG, "ServiceAbility::onStop");
28      }
29
30      @Override
31      public void onCommand(Intent intent, boolean restart, int startId) {
32      }
33
34      @Override
35      public IRemoteObject onConnect(Intent intent) {
36          return null;
37      }
38
39      @Override
40      public void onDisconnect(Intent intent) {
41      }
42  }
```

- onStart()：该方法在创建 Service Ability 的时候调用，用于 Service Ability 的初始化。在 Service Ability 的整个生命周期只会调用一次，调用时传入的 Intent 应为空。
- onCommand()：在 Service Ability 创建完成之后调用，该方法在客户端每次启动该 Service Ability 时都会调用，用户可以在该方法中做一些调用统计、初始化类的操作。
- onConnect()：在 Ability 和 Service Ability 连接时调用，该方法返回 IRemoteObject

对象,用户可以在该回调函数中生成对应 Service Ability 的 IPC 通信通道,以便 Ability 与 Service Ability 交互。Ability 可以多次连接同一个 Service Ability,系统会缓存该 Service Ability 的 IPC 通信对象,只有第一个客户端连接 Service Ability 时,系统才会调用 Service Ability 的 onConnect() 方法生成 IRemoteObject 对象,之后系统会将同一个 RemoteObject 对象传递至其他连接同一个 Service Ability 的所有客户端,无须再次调用 onConnect 方法。

- onDisconnect():在 Ability 与绑定的 Service Ability 断开连接时调用。
- onStop():在 Service Ability 销毁时调用。Service Ability 应通过实现此方法来清理任何资源,如关闭线程、注册的侦听器等。
- onBackground():在 Service Ability 后台运行时调用。

4.3.2 启动 Service Ability

Ability 为开发者提供了 startAbility() 方法来启动另外一个 Ability。因为 Service Ability 也是 Ability 的一种,开发者同样可以通过将 Intent 传递给该方法来启动 Service Ability。不仅支持启动本地 Service Ability,还支持启动远程 Service Ability。

开发者可以通过构造包含 DeviceId、BundleName 与 AbilityName 的 Operation 对象来设置目标 Service 信息。

- DeviceId:表示设备 ID。如果是本地设备,可以直接留空;如果是远程设备,可以通过 ohos.distributedschedule.interwork.DeviceManager 提供的 getDeviceList() 方法获取设备列表。
- BundleName:表示包名称。
- AbilityName:表示待启动的 Ability 名称。

下面用代码实践。比如我现在要在 MainAbilitySlice 的 onStart() 方法中启动 ServiceAbility。启动本地设备 Service Ability 的详细代码如程序清单 4-24 所示。

程序清单 4-24:hmos\ch04\01\AbilityTest\entry\src\main\java\com\example\abilitytest\slice\MainAbilitySlice.java

```
1    private void startupLocalService() {
2        Intent intent = new Intent();
3        Operation operation = new Intent.OperationBuilder()  //构建操作方式
4            .withDeviceId("")                    //设备 id,传入空字符串表示为当前设备
5            .withBundleName("com.example.abilitytest")       //应用的包名
6            .withAbilityName("com.example.abilitytest.ServiceAbility")
                                                  //跳转目标的路径名
7            .build();
8        intent.setOperation(operation);          //设置操作
9        startAbility(intent);
10   }
```

然后在 onStart() 中调用即可,详细代码如程序清单 4-25 所示。

程序清单 4-25：hmos\ch04\01\AbilityTest\entry\src\main\java\
com\example\abilitytest\slice\MainAbilitySlice.java

```
11    @Override
12    public void onStart(Intent intent) {
13        super.onStart(intent);
14        //加载 XML 布局
15        super.setUIContent(ResourceTable.Layout_ability_main);
16        ...
17        //启动本地设备 Service
18        startupLocalService();
19        ...
20    }
```

那么，怎么证明 Service Ability 启动了？很简单，只要在 ServiceAbility 的 onStart() 方法中打印一个日志就可以了（日志的使用方法在本书 6.4 节中有详细介绍）。打开 ServiceAbility.java，会发现在创建 Service Ability 时就已经创建好了日志，如图 4-20 所示。

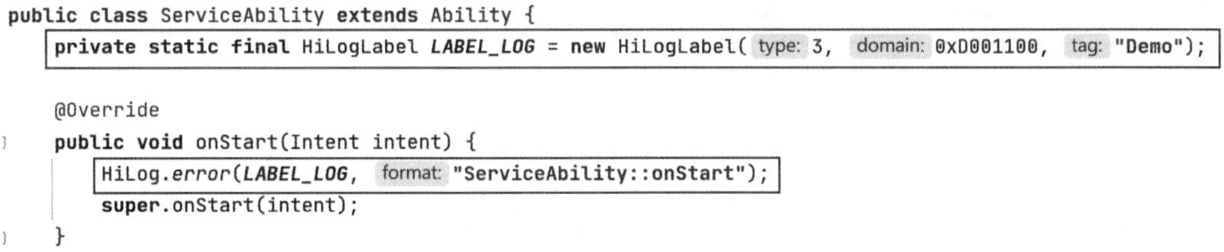

图 4-20　Service Ability 中的日志一

现在启动模拟器运行项目。项目运行成功后，Service Ability 也一起启动，当出现如图 4-21 所示的日志输出时，表示 Service Ability 启动成功。

图 4-21　Service Ability 中的日志输出一

启动远程设备中的 Service Ability 的方法和启动本地设备的 Service Ability 类似，只需要在构建 Operation 对象时设置已连接的设备 ID，已连接的远程设备列表可以通过 DeviceManager 的 getDeviceList() 方法获取，设备 ID 可以通过 getDeviceId() 得到。启动远程设备 Service Ability 的详细代码如程序清单 4-26 所示。

程序清单 4-26 hmos\ch04\01\AbilityTestVentry\src\main\java\com\example\abilitytest\slice\MainAbilitySlice.java

```java
1   private void startRemoteService() {
2       Intent intent = new Intent();
3       //获取远程设备列表
4       List<DeviceInfo> deviceInfos = DeviceManager.getDeviceList(DeviceInfo.FLAG_GET_ONLINE_DEVICE);
5       for (DeviceInfo deviceInfo : deviceInfos) {
6           Operation operation = new Intent.OperationBuilder() //构建操作方式
7                   .withDeviceId(deviceInfo.getDeviceId()) //设置远程设备的 ID
8                   .withBundleName("com.example.abilitytest")  //应用的包名
9                   .withAbilityName("com.example.abilitytest.ServiceAbility")
                                                                //应用的包名
10                  .withFlags(Intent.FLAG_ABILITYSLICE_MULTI_DEVICE)
                                //设置支持分布式调度系统多设备启动的标识
11                  .build();
12          intent.setOperation(operation);                     //设置操作
13          startAbility(intent);
14      }
15  }
```

远程启动 Service Ability 时，还需要设置启动的这个 Service Ability 允许其他应用程序发现。我们需要到配置文件（config.json）中找到 HUAWEI DevEco Studio 为 Service Ability 配置的相关代码，添加一个 visible 属性，并将其设置为 true，如果没有设置这个属性，它的值默认为 false。具体代码如图 4-22 所示。

```
{
  "visible": true,
  "name": "com.example.abilitytest.ServiceAbility",
  "icon": "$media:icon",
  "description": "$string:serviceability_description",
  "type": "service"
}
```

图 4-22 远程启动 Service Ability 所需的配置信息

执行上述代码后，Ability 将通过 startAbility()方法启动 Service Ability。如果 Service Ability 尚未运行，则系统会先调用 onStart()初始化 Service Ability，再回调 Service Ability 的 onCommand()方法启动 Service Ability。刚才我们并没有在系统回调 onCommand()方法时看到日志输出，是因为它里面没有相应的日志输出语句。现在在 onCommand()方法中也加上日志输出语句，详细代码如程序清单 4-27 所示。

程序清单 4-27：hmos\ch04\01\AbilityTest\entry\src\main\
java\com\example\abilitytest\ServiceAbility.java

```
1  @Override
2  public void onCommand(Intent intent, boolean restart, int startId) {
3      HiLog.error(LABEL_LOG, "ServiceAbility::onCommand");
4  }
```

然后重新运行，结果如图 4-23 所示。

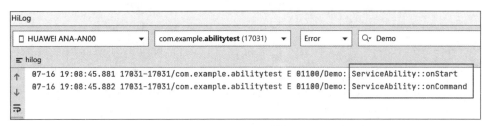

图 4-23　Service Ability 中的日志输出二

如果 Service Ability 正在运行，则系统会直接回调 Service Ability 的 onCommand()方法来启动 Service Ability。这个场景需要先返回到设备主页面，然后再打开这个应用，首先返回主页面，单击中间的圆形按钮，如图 4-24(a)所示。这时 Service Ability 在后台运行，我们再次打开应用，如图 4-24(b)和图 4-24(c)所示。

(a) 单击返回主界面按钮　　　(b) 单击应用程序　　　(c) 回到应用程序

图 4-24　Service Ability 案例

此时，再查看日志信息，如图 4-25 所示。系统直接回调 Service Ability 的 onCommand()方法来启动 Service，并没有先调用 onStart()来初始化 Service Ability。

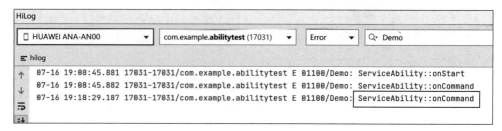

图 4-25　Service Ability 中的日志输出三

4.3.3　停止 Service Ability

Service Ability 一旦创建，就会一直保持在后台运行，除非必须回收内存资源，否则系统不会停止或销毁 Service Ability。开发者可以在其他 Ability 调用 stopAbility()来停止 Service Ability 或在 Service Ability 中通过 terminateAbility()停止本 Service Ability。停止 Service Ability 同样支持停止本地设备 Service Ability 和停止远程设备 Service Ability，使用方法与启动 Service Ability 一样。一旦调用停止 Service Ability 的方法，系统便会销毁 Service Ability。

先尝试第一种停止 Service Ability 的方法：在 Page Ability 中停止。首先在 MainAbilitySlice 中增加一个停止服务的方法，详细代码如程序清单 4-28 所示。

程序清单 4-28：hmos\ch04\01\AbilityTest\entry\src\main\java\
com\example\abilitytest\slice\MainAbilitySlice.java

```
1   private void stopLocalService() {
2       Intent intent = new Intent();
3       Operation operation = new Intent.OperationBuilder()    //构建操作方式
4           .withDeviceId("")                                  //设备 ID
5           .withBundleName("com.example.abilitytest")         //应用的包名
6           .withAbilityName("com.example.abilitytest.ServiceAbility")
                                                               //跳转目标的路径名
7           .build();
8       intent.setOperation(operation);                        //设置操作
9       stopAbility(intent);                                   //停止服务
10  }
```

然后在 MainAbilitySlice 的 onStart()中添加单击事件用来停止 Service Ability，详细代码如程序清单 4-29 所示。

程序清单 4-29：hmos\ch04\01\AbilityTest\entry\src\main\java\
com\example\abilitytest\slice\MainAbilitySlice.java

```
1   @Override
```

```
2   public void onStart(Intent intent) {
3       super.onStart(intent);
4       //加载 XML 布局
5       super.setUIContent(ResourceTable.Layout_ability_main);
6       //启动本地设备 Service
7       startupLocalService();
8       //设置单击事件
9       Button button_stop = (Button) findComponentById(ResourceTable.Id_
    button_stop);
10      button_stop.setClickedListener(new Component.ClickedListener() {
11          @Override
12          public void onClick(Component component) {
13              stopLocalService();              //停止本地 Service Ability
14          }
15      });
16  }
```

接着,先运行项目进入主页面,单击 STOP SERVICE 按钮查看日志信息,图 4-26 所示表示停止成功。

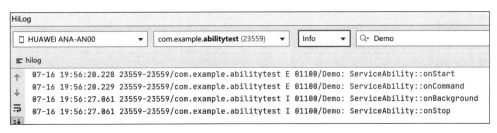

图 4-26　Service Ability 中的日志输出四

从此可以看出,当从其他的 Page Ability 中停止 Service Ability 时,会先回调 onBackground(),因为这个时候 Service Ability 是在前台运行的,系统会把 Service Ability 放到后台,然后再停止这个服务。

下面尝试第二种停止 Service Ability 的方法:在 Service Ability 中停止。可以通过一个延时服务来操作,详细代码如程序清单 4-30 所示。

程序清单 4-30　hmos\ch04\01\AbilityTest\entry\src\main\java\
com\example\abilitytest\slice\MainAbilitySlice.java

```
1   ScheduledExecutorService scheduledExecutorService = Executors.
    newScheduledThreadPool(1);
2
3   private void stopService() {
4       //延时任务
5       scheduledExecutorService.schedule(new Runnable() {
```

```
6        @Override
7        public void run() {
8            terminateAbility();              //停止当前服务
9        }
10   }, 3, TimeUnit.SECONDS);                  //延时 3s 执行
11 }
```

在 onStart()中调用 stopService()将项目运行到远程模拟机上,然后等待 3s 就会自动调用 terminateAbility()停止 Service Ability,可以发现此方法和第一种方法在其他 Page Ability 中停止的执行流程是一样的,图 4-27 所示表示自动停止成功,其中连接成功到停止成功相隔 3s。

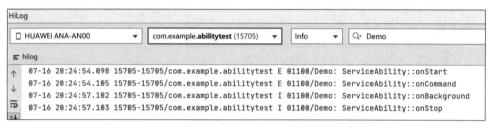

图 4-27　Service Ability 中的日志输出五

4.3.4　连接 Service Ability

如果 Service Ability 需要与 Page Ability 或其他应用的 Service Ability 进行交互,则须创建用于连接的 Connection。Service Ability 支持其他 Ability 通过 connectAbility() 方法与其进行连接。

在使用 connectAbility()处理回调时,需要传入目标 Service Ability 的 Intent 与 IAbilityConnection 的实例。IAbilityConnection 提供了两个方法供开发者实现:onAbility-ConnectDone()是用来处理连接 Service Ability 成功的回调;onAbilityDisconnectDone() 是用来处理 Service Ability 异常死亡的回调。

下面通过一个示例学习具体的连接方法。首先,在 MainAbilitySlice 中添加如程序清单 4-31 所示的代码。

程序清单 4-31:hmos\ch04\01\AbilityTest\entry\src\main\java\
com\example\abilitytest\slice\MainAbilitySlice.java

```
1  //创建连接 Service Ability 的回调实例
2  private IAbilityConnection iAbilityConnection = new IAbilityConnection() {
3      //连接 Service Ability 成功的回调
4      @Override
5      public void onAbilityConnectDone(ElementName elementName,
   IRemoteObject iRemoteObject, int resultCode) {
```

```
6              //在这里开发者可以拿到服务端传过来的IRemoteObject对象,从中解析出
   //服务端传过来的信息
7          }
8
9      // Service Ability 异常死亡的回调
10     @Override
11     public void onAbilityDisconnectDone(ElementName elementName, int
   resultCode) {
12         }
13 };
14
15 //连接 Service Ability
16 private void connectService() {
17     Intent intent = new Intent();
18     Operation operation = new Intent.OperationBuilder()   //构建操作方式
19         .withDeviceId("")                                  //设备 ID
20         .withBundleName("com.example.abilitytest")        //应用的包名
21         .withAbilityName("com.example.abilitytest.ServiceAbility")
                                                             //跳转目标的路径名
22         .build();
23     intent.setOperation(operation);                       //设置操作
24     connectAbility(intent, iAbilityConnection);           //连接到服务
25 }
```

在上述代码中,onAbilityConnectDone()是用来处理连接 Service Ability 成功的回调,其中 elementName 为连接设备的相关信息,IRemoteObject 相当于 Service Ability 连接通道中的数据包,resultCode 是返回结果码,值为 0 代表连接成功,否则为连接失败。onAbilityDisconnectDone()是用来处理连接 Service Ability 异常死亡的回调,其中各参数的意义与 onAbilityConnectDone()中的一致。

然后在 MainAbilitySlice 的 onStart() 中添加单击事件,在单击按钮时调用 connectService()来连接 Service Ability,详细代码如程序清单 4-32 所示。

程序清单 4-32:hmos\ch04\01\AbilityTest\entry\src\main\java\
com\example\abilitytest\slice\MainAbilitySlice.java

```
1  @Override
2  public void onStart(Intent intent) {
3      super.onStart(intent);
4      //加载 XML 布局
5      super.setUIContent(ResourceTable.Layout_ability_main);
6      //启动本地设备 Service
7      startupLocalService();
8      //设置单击事件
```

```
9       ...
10      Button button_connection = (Button) findComponentById(ResourceTable.
Id_button_connection);
11      button_connection.setClickedListener(new Component.ClickedListener() {
12          @Override
13          public void onClick(Component component) {
14              connectService();
15          }
16      });
17  }
```

同时,Service Ability 侧也需要在 onConnect()时返回 IRemoteObject,从而定义与 Service Ability 进行通信的接口。HarmonyOS 提供了 IRemoteObject 的默认实现,用户可以通过继承 LocalRemoteObject 来创建自定义的实现类,也可以直接继承 RemoteObject。并且在 onConnect()方法中加入日志打印,以便观察其生命周期,详细代码如程序清单 4-33 所示。

程序清单 4-33: hmos\ch04\01\AbilityTest\entry\src\main\java\com\example\abilitytest\ServiceAbility.java

```
1   //把 IRemoteObject 返回给客户端
2   @Override
3   public IRemoteObject onConnect(Intent intent) {
4       HiLog.error(LABEL_LOG, "ServiceAbility::onConnect");
5       return new MyRemoteObject();
6   }
7
8   //创建自定义 IRemoteObject 实现类
9   private class MyRemoteObject extends LocalRemoteObject {
10      private MyRemoteObject() {
11          super();
12      }
13  }
```

把 4.3.3 节中在 Service Ability 的 onStart()方法中添加的用于停止 Service Ability 的方法 stopService()注释掉后运行项目,然后单击 CONNECTION SERVICE 按钮查看日志信息,图 4-28 所示表示连接成功。

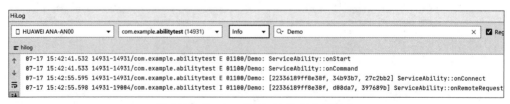

图 4-28 Service Ability 中的日志输出六

4.3.5 断开 Service Ability

断开服务其实就比较简单了,通过调用 disconnectAbility()方法,把之前创建的 Service Ability 回调实例传入即可。同样,在 MainAbilitySlice 中创建一个延时任务并在之前创建的 Service Ability 回调实例中的 onAbilityConnectDone()中调用,让 Service Ability 在连接成功的 3s 后断开连接,详细代码如程序清单 4-34 所示。

程序清单 4-34:hmos\ch04\01\AbilityTest\entry\src\main\java\com\example\abilitytest\slice\MainAbilitySlice.java

```
1   //创建连接Service的回调实例
2   private IAbilityConnection iAbilityConnection = new IAbilityConnection() {
3       @Override
4       public void onAbilityConnectDone(ElementName elementName, IRemoteObject iRemoteObject, int resultCode) {
5           //在这里开发者可以拿到服务端传过来的IRemoteObject对象,从中解析出
            //服务端传过来的信息
6           stopService();                                      //3s后断开连接
7       }
8
9       // Service 异常死亡的回调
10      @Override
11      public void onAbilityDisconnectDone(ElementName elementName, int resultCode) {
12      }
13  };
14
15  //创建一个线程池
16  ScheduledExecutorService scheduledExecutorService = Executors.newScheduledThreadPool(1);
17
18  private void stopService() {
19      //延时任务,3s后断开连接
20      scheduledExecutorService.schedule(new Runnable() {
21          @Override
22          public void run() {
23              disconnectAbility(iAbilityConnection);           //断开连接
24          }
25      }, 3, TimeUnit.SECONDS);                                 //延时3s执行
26  }
```

在 ServiceAbility 的 onDisconnect()方法中添加日志打印,以便观察其生命周期,详细代码如程序清单 4-35 所示。

程序清单 4-35：hmos\ch04\01\AbilityTest\entry\src\main\
java\com\example\abilitytest\ServiceAbility.java

```
1   @Override
2   public void onDisconnect(Intent intent) {
3       HiLog.error(LABEL_LOG, "ServiceAbility::onDisconnect");
4   }
```

最后，运行项目，单击 CONNECTION SERVICE 按钮查看日志信息，图 4-29 所示表示断开连接成功，其中由连接成功到断开连接相隔 3s。

图 4-29 Service Ability 中的日志输出七

4.3.6 利用 Service Ability 处理数据

这里简单模拟客户端传递一个 int 类型的数据给服务端，然后服务端将数据乘 2 再返回。

首先新建一个类或者在 MainAbilitySlice 中创建客户端代理类并实现 IRemoteBroker 接口，IRemoteBroker 翻译过来是远程经纪人的意思，这就像两个球员之间通过经纪人来交流，而并非两个球员直接交流。这里我们直接在 MainAbilitySlice 中创建，并在里面创建 sendMessageToService() 方法用来向远程发布消息，详细代码如程序清单 4-36 所示。

程序清单 4-36：hmos\ch04\01\AbilityTest\entry\src\main\java\
com\example\abilitytest\slice\MainAbilitySlice.java

```
1   private class MyIRemoteBroker implements IRemoteBroker {
2       private static final int SUCCESS = 0;
3       private static final int OPERATION= 1;
4       private final IRemoteObject iRemoteObject;
5
6       public MyIRemoteBroker (IRemoteObject iRemoteObject) {
7           this.iRemoteObject= iRemoteObject;
8       }
9
10      public int sendMessageToService(int num) {
11          MessageParcel data = MessageParcel.obtain();
12          data.writeInt(num);
```

```
13          MessageParcel reply = MessageParcel.obtain();
14          MessageOption option = new MessageOption(MessageOption.TF_SYNC);
15          int result = 0;
16          try {
17              iRemoteObject.sendRequest(OPERATION, data, reply, option);
18              int resultCode = reply.readInt();
19              if (resultCode != SUCCESS) {
20                  throw new RemoteException();
21              }
22              result = reply.readInt();
23          } catch (RemoteException e) {
24              e.printStackTrace();
25          }
26          return result;
27      }
28
29      @Override
30      public IRemoteObject asObject() {
31          return iRemoteObject;
32      }
33  }
```

上述代码中，具体用来向远程发布消息的代码是 iRemoteObject.sendRequest（OPERATION，data，reply，option），其中第一个参数为请求码，由两端约定好，当客户端发送某一请求码时，服务端就通过识别该请求码判断接下来要进行哪种操作。第二和第三个参数传递的均为 MessageParcel 类型，MessageParcel 类继承自 Parcel 类，使用的方法和 Android 中的 Parcel 类类似，其中第二个参数是客户端向服务端传递的消息，第三个参数是服务端向客户端回复的消息，在发送请求时就已经把需要接收的消息的位置留好了，是为了让服务端接收数据后可以不用将数据清空后再写入结果，使得请求和结果互不干扰。第四个参数是设定本次通信是同步操作还是异步操作，如需设定为同步操作，则在初始化时传入 MessageOption.TF_SYNC；如需设定为异步操作，则传入 MessageOption.TF_ASYNC。

接下来，在之前创建的 Service Ability 回调实例中使用 MyIRemoteBroker，并在 onStart()方法中添加按钮的事件监听，以便在单击按钮后向服务端发送请求并在日志中打印出处理结果（因为此处的按钮不需要在其他地方接着使用，所以我们使用另一种方式设置监听事件），这里传入数值 512 到服务端，详细代码如程序清单 4-37 所示。

程序清单 4-37：hmos\ch04\01\AbilityTest\entry\src\main\java\com\example\abilitytest\slice\MainAbilitySlice.java

```
1   private IAbilityConnection iAbilityConnection = new IAbilityConnection() {
2       @Override
```

```
3    public void onAbilityConnectDone(ElementName elementName,
     IRemoteObject iRemoteObject, int resultCode) {
4        myIRemoteBroker = new MyIRemoteBroker(iRemoteObject);
5        HiLog.info(LABEL_LOG, "MainAbilitySlice::
     onAbilityConnectDone");
6    }
7
8    @Override
9    public void onAbilityDisconnectDone(ElementName elementName, int
     resultCode) {
10       HiLog.info(LABEL_LOG, "MainAbilitySlice::
     onAbilityDisconnectDone");
11   }
12 };
13
14 @Override
15 public void onStart(Intent intent) {
16     super.onStart(intent);
17     super.setUIContent(ResourceTable.Layout_ability_main);
18     startupLocalService();
19     ...
20     findComponentById(ResourceTable.Id_button_test).setClickedListener
     (new Component.ClickedListener() {
21         @Override
22         public void onClick(Component component) {
23             if (myIRemoteBroker != null) {
24                 int result = myIRemoteBroker.sendMessageToService(512);
                                                     //传入 512 交给服务器处理
25                 HiLog.info(LABEL_LOG, "result = " + result);
26             } else {
27                 HiLog.error(LABEL_LOG, "clientRemoteProxy != null");
28             }
29         }
30     });
31 }
```

把数据传输到服务端后,就是对数据进行处理再回传回去,这里我们重写过之前在 ServiceAbility 中创建的 MyRemoteObject 类,让它继承 RemoteObject 并实现 IRemoteBroker 接口,然后在 onConnect() 中返回 MyRemoteObject 实例,详细代码如程序清单 4-38 所示。

```
1   //创建自定义 IRemoteObject 实现类
2   private class MyRemoteObject extends RemoteObject implements
    IRemoteBroker {
3       private static final int SUCCESS = 0;
4       private static final int FAILED = -1;
5       private static final int OPERATION = 1;
6
7       public MyRemoteObject(String descriptor) {
8           super(descriptor);
9       }
10
11      @Override
12      public IRemoteObject asObject() {
13          return this;
14      }
15      //接收客户端发送的请求
16      @Override
17      public boolean onRemoteRequest(int code, MessageParcel data,
    MessageParcel reply, MessageOption option) throws RemoteException {
18          if (code != OPERATION) {
19              reply.writeInt(FAILED);
20              return false;
21          }
22          int initData = data.readInt();              //读取客户端传入的数据
23          int resultData = initData * 2;              //将客户端传入的数据乘 2
24          reply.writeInt(SUCCESS);
25          reply.writeInt(resultData);
26          return true;
27      }
28  }
29
30  //把 IRemoteObject 返回给客户端
31  @Override
32  public IRemoteObject onConnect(Intent intent) {
33      HiLog.error(LABEL_LOG, "ServiceAbility::onConnect");
34      return new MyRemoteObject("");
35  }
```

所有代码完成后,运行项目,先单击 CONNECTION SERVICE 按钮连接 Service Ability,再单击 TEST 按钮模拟客户端传第一个参数给服务端处理后返回,然后打开

HiLog 选项卡查看日志，如图 4-30 所示。

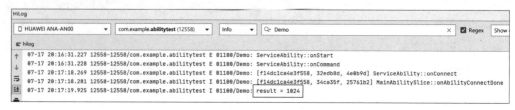

图 4-30 Service Ability 中的日志输出八

4.3.7 前台 Service Ability

通常情况下，如果开发者不作特别的设置，Service Ability 都是在后台运行的，然而后台 Service Ability 的优先级比较低，当资源不足时，系统可能会回收正在运行的后台 Service Ability。但在某些场景下，用户希望应用程序的某些功能能够一直保持运行，此时开发者就需要使用到前台 Service Ability，前台 Service Ability 会始终保持正在运行的图标在系统状态栏显示。

前台 Service Ability 的开发并不复杂，开发者只需在配置文件中声明 ohos.permission.KEEP_BACKGROUND_RUNNING 权限以及添加对应的 backgroundModes 参数后在 Service Ability 创建的方法里调用 keepBackgroundRunning()将 Service Ability 与通知进行绑定就可以了。在 onStop()方法中调用 cancelBackgroundRunning()方法可停止前台 Service Ability。

下面具体讲解开发前台 Service Ability 的核心代码和所需的配置。首先，先在 Service Ability 中新建一个启动前台服务的方法，然后在 onStart()中调用该方法，详细代码如程序清单 4-39 所示。

程序清单 4-39：hmos\ch04\01\AbilityTest\entry\src\main\java\com\example\abilitytest\ServiceAbility.java

```
1   @Override
2   public void onStart(Intent intent) {
3       HiLog.error(LABEL_LOG, "ServiceAbility::onStart");
4       super.onStart(intent);
5       ...
6       startupForegroundService();
7   }
8
9   //启动前台服务
10  private void startupForegroundService() {
11      NotificationRequest request = new NotificationRequest(1005);
12      NotificationRequest.NotificationNormalContent content = new NotificationRequest.NotificationNormalContent();
13      content.setTitle("title").setText("text");
```

```
14      NotificationRequest.NotificationContent notificationContent = new
   NotificationRequest.NotificationContent(content);
15      request.setContent(notificationContent);
16      //绑定通知,1005 为创建通知时传入的 notificationId
17      keepBackgroundRunning(1005, request);
18  }
```

之后在配置文件(config.json)中的 module＞abilities 字段下对当前 Service Ability 进行如图 4-31 所示的配置。

```
{
  "visible": true,
  "name": "com.example.abilitytest.ServiceAbility",
  "icon": "$media:icon",
  "description": "hap sample empty service",
  "type": "service",
  "backgroundModes": [
    "dataTransfer"
  ],
  "permissions": [
    "ohos.permission.KEEP_BACKGROUND_RUNNING"
  ]
}
```

图 4-31　前台 Service Ability 所需的配置信息

4.3.8　示例的完整代码

在 AbilityTest 案例工程中有关 Service Ability 的关键文件 MainAbilitySlice.java 和 ServiceAbility.java 的完整代码如程序清单 4-40 和程序清单 4-41 所示。

程序清单 4-40：hmos\ch04\01\AbilityTest\entry\src\main\java\com\example\abilitytest\slice\MainAbilitySlice.java

```
1   public class MainAbilitySlice extends AbilitySlice {
2       private static final HiLogLabel LABEL_LOG = new HiLogLabel(3,
    0xD001100, "Demo");
3       MyIRemoteBroker myIRemoteBroker = null;
4
5       @Override
6       public void onStart(Intent intent) {
7           super.onStart(intent);
8           super.setUIContent(ResourceTable.Layout_ability_main);
                                                            //加载 XML 布局
9           startupLocalService();          //启动本地设备 Service
10          //设置单击事件
```

```java
11          Button button_stop = (Button) findComponentById(ResourceTable.Id_button_stop);
12          button_stop.setClickedListener(new Component.ClickedListener() {
13              @Override
14              public void onClick(Component component) {
15                  stopLocalService();                  //停止本地 Service Ability
16              }
17          });
18          Button button_connection = (Button) findComponentById(ResourceTable.Id_button_connection);
19          button_connection.setClickedListener(new Component.ClickedListener() {
20              @Override
21              public void onClick(Component component) {
22                  connectService();
23              }
24          });
25          findComponentById(ResourceTable.Id_button_test).setClickedListener(new Component.ClickedListener() {
26              @Override
27              public void onClick(Component component) {
28                  if (myIRemoteBroker != null) {
29                      int result = myIRemoteBroker.sendMessageToService(512);
                                                                     //传入 512 交给服务端处理
30                      HiLog.info(LABEL_LOG, "result = " + result);
31                  } else {
32                      HiLog.error(LABEL_LOG, "clientRemoteProxy != null");
33                  }
34              }
35          });
36      }
37
38      //启动本地设备 Service Ability
39      private void startupLocalService() {
40          Intent intent = new Intent();
41          Operation operation = new Intent.OperationBuilder()   //构建操作方式
42                  .withDeviceId("")         //设备 ID,传入空字符串表示为当前设备
43                  .withBundleName("com.example.abilitytest")   //应用的包名
44                  .withAbilityName("com.example.abilitytest.ServiceAbility")
                                                                 //跳转目标的路径名
45                  .build();
46          intent.setOperation(operation);                      //设置操作
47          startAbility(intent);
48      }
```

```
49
50        //启动远程设备 Service Ability
51        private void startRemoteService() {
52            Intent intent = new Intent();
53            //获取远程设备列表
54            List<DeviceInfo> deviceInfos = DeviceManager.getDeviceList
   (DeviceInfo.FLAG_GET_ONLINE_DEVICE);
55            for (DeviceInfo deviceInfo : deviceInfos) {
56                Operation operation = new Intent.OperationBuilder()
                                                                                //构建操作方式
57                        .withDeviceId(deviceInfo.getDeviceId())   //设备 ID
58                        .withBundleName("com.example.abilitytest")    //应用的包名
59                        .withAbilityName("com.example.abilitytest.ServiceAbility")
                                                                                //跳转目标的路径名
60                        .withFlags(Intent.FLAG_ABILITYSLICE_MULTI_DEVICE)
                                                        //设置支持分布式调度系统多设备启动的标识
61                        .build();
62                intent.setOperation(operation);                               //设置操作
63                startAbility(intent);
64            }
65        }
66
67        //在 Page Ability 中停止本地 Service Ability
68        private void stopLocalService() {
69            Intent intent = new Intent();
70            Operation operation = new Intent.OperationBuilder()  //构建操作方式
71                    .withDeviceId("")                             //设备 ID
72                    .withBundleName("com.example.abilitytest")    //应用的包名
73                    .withAbilityName("com.example.abilitytest.ServiceAbility")
                                                                    //跳转目标的路径名
74                    .build();
75            intent.setOperation(operation);                       //设置操作
76            stopAbility(intent);                                  //停止服务
77        }
78
79        //创建连接 Service Ability 回调实例
80        private IAbilityConnection iAbilityConnection = new
   IAbilityConnection() {
81            //连接 Service Ability 成功的回调
82            @Override
83            public void onAbilityConnectDone(ElementName elementName,
   IRemoteObject iRemoteObject, int resultCode) {
84                //在这里开发者可以拿到服务端传过来的 IRemoteObject 对象,从中解析
   //出服务端传过来的信息
```

```
85    //        stopService();
86            myIRemoteBroker = new MyIRemoteBroker(iRemoteObject);
87            HiLog.info(LABEL_LOG, "MainAbilitySlice::onAbilityConnectDone");
88        }
89
90        // Service 异常死亡的回调
91        @Override
92        public void onAbilityDisconnectDone(ElementName elementName, int resultCode) {
93            HiLog.info(LABEL_LOG, "MainAbilitySlice::onAbilityDisconnectDone");
94        }
95    };
96
97    //连接 Service Ability
98    private void connectService() {
99        Intent intent = new Intent();
100       Operation operation = new Intent.OperationBuilder()                //构建操作方式
101               .withDeviceId("")                                           //设备 ID
102               .withBundleName("com.example.abilitytest")                  //应用的包名
103               .withAbilityName("com.example.abilitytest.ServiceAbility")  //跳转目标的路径名
104               .build();
105       intent.setOperation(operation);                                    //设置操作
106       connectAbility(intent, iAbilityConnection);                        //连接到服务
107   }
108
109   //创建一个线程池
110   ScheduledExecutorService scheduledExecutorService = Executors.newScheduledThreadPool(1);
111
112   private void stopService() {
113       //延时任务
114       scheduledExecutorService.schedule(new Runnable() {
115           @Override
116           public void run() {
117               disconnectAbility(iAbilityConnection);                     //断开连接
118           }
119       }, 3, TimeUnit.SECONDS);                                           //延时 3s 执行
120   }
121
```

```java
122    public class MyIRemoteBroker implements IRemoteBroker {
123        private static final int SUCCESS = 0;
124        private static final int OPERATION= 1;
125        private final IRemoteObject iRemoteObject;
126
127        public MyIRemoteBroker(IRemoteObject iRemoteObject) {
128            this.iRemoteObject = iRemoteObject;
129        }
130
131        public int sendMessageToService(int num) {
132            MessageParcel data = MessageParcel.obtain();
133            data.writeInt(num);
134            MessageParcel reply = MessageParcel.obtain();
135            MessageOption option = new MessageOption(MessageOption.TF_SYNC);
136            int result = 0;
137            try {
138                iRemoteObject.sendRequest(OPERATION, data, reply, option);
139                int resultCode = reply.readInt();
140                if (resultCode != SUCCESS) {
141                    throw new RemoteException();
142                }
143                result = reply.readInt();
144            } catch (RemoteException e) {
145                e.printStackTrace();
146            }
147            return result;
148        }
149
150        @Override
151        public IRemoteObject asObject() {
152            return iRemoteObject;
153        }
154    }
155
156    @Override
157    public void onActive() {
158        super.onActive();
159    }
160
161    @Override
162    public void onForeground(Intent intent) {
163        super.onForeground(intent);
164    }
165 }
```

程序清单 4-41：hmos\ch04\01\AbilityTest\entry\src\main\java\
com\example\abilitytest\ServiceAbility.java

```java
1   public class ServiceAbility extends Ability {
2       private static final HiLogLabel LABEL_LOG = new HiLogLabel(3, 0xD001100, "Demo");
3
4       @Override
5       public void onStart(Intent intent) {
6           HiLog.error(LABEL_LOG, "ServiceAbility::onStart");
7           super.onStart(intent);
8   //        stopService();//停止 Service Ability
9   //        startupForegroundService();
10      }
11
12      @Override
13      public void onBackground() {
14          super.onBackground();
15          HiLog.info(LABEL_LOG, "ServiceAbility::onBackground");
16      }
17
18      @Override
19      public void onStop() {
20          super.onStop();
21          HiLog.info(LABEL_LOG, "ServiceAbility::onStop");
22      }
23
24      @Override
25      public void onCommand(Intent intent, boolean restart, int startId) {
26          HiLog.error(LABEL_LOG, "ServiceAbility::onCommand");
27      }
28
29      //把 IRemoteObject 返回给客户端
30      @Override
31      public IRemoteObject onConnect(Intent intent) {
32          HiLog.error(LABEL_LOG, "ServiceAbility::onConnect");
33          return new MyRemoteObject("");
34      }
35
36      @Override
37      public void onDisconnect(Intent intent) {
38          HiLog.error(LABEL_LOG, "ServiceAbility::onDisconnect");
39      }
```

```
40
41        //创建一个线程池
42        ScheduledExecutorService scheduledExecutorService = Executors.
   newScheduledThreadPool(1);
43
44        private void stopService() {
45            //延时任务
46            scheduledExecutorService.schedule(new Runnable() {
47                @Override
48                public void run() {
49                    terminateAbility();              //停止当前服务
50                }
51            }, 3, TimeUnit.SECONDS);                  //延时 3s 执行
52        }
53
54        //创建自定义 IRemoteObject 实现类
55        private class MyRemoteObject extends RemoteObject implements
   IRemoteBroker {
56            private static final int SUCCESS = 0;
57            private static final int FAILED = -1;
58            private static final int OPERATION = 1;
59
60            public MyRemoteObject(String descriptor) {
61                super(descriptor);
62            }
63
64            @Override
65            public IRemoteObject asObject() {
66                return this;
67            }
68            //接收客户端发送的请求
69            @Override
70            public boolean onRemoteRequest(int code, MessageParcel data,
   MessageParcel reply, MessageOption option) throws RemoteException {
71                if (code != OPERATION) {
72                    reply.writeInt(FAILED);
73                    return false;
74                }
75                int initData = data.readInt();    //读取客户端传入的数据
76                int resultData = initData * 2;    //将客户端传入的数据乘 2
77                reply.writeInt(SUCCESS);
78                reply.writeInt(resultData);
79                return true;
```

```
80          }
81      }
82
83      //启动前台服务
84      private void startupForegroundService() {
85          NotificationRequest request = new NotificationRequest(1005);
86          NotificationRequest.NotificationNormalContent content = new NotificationRequest.NotificationNormalContent();
87          content.setTitle("title").setText("text");
88          NotificationRequest.NotificationContent notificationContent = new NotificationRequest.NotificationContent(content);
89          request.setContent(notificationContent);
90          //绑定通知,1005为创建通知时传入的notificationId
91          keepBackgroundRunning(1005, request);
92      }
93  }
```

4.4 Data Ability

一个应用通常会有许多数据需要存储或访问,开发者以往需要建立一个便捷高效的工具类来处理这些操作,这样对于开发者来说是很麻烦的。而 HarmonyOS 为开发者提供了 Data Ability 用于解决这个问题,Data Ability 有助于应用管理其自身和其他应用存储数据的访问,并提供与其他应用共享数据的方法。Data Ability 既可用于同设备不同应用的数据共享,也支持跨设备不同应用的数据共享。Data Ability 对外提供对数据的增加、删除、修改、查询等接口,但这些接口的具体实现需由开发者提供。

4.4.1 URI 介绍

Data Ability 的提供方和使用方都通过 URI(Uniform Resource Identifier)标识一个具体的数据,例如数据库中的某个表或磁盘上的某个文件。HarmonyOS 的 URI 仍基于 URI 通用标准,格式如图 4-32 所示。

图 4-32 URI 格式

- scheme:协议方案名,固定为 dataability,代表 Data Ability 使用的协议类型。
- authority:设备 ID。如果为跨设备场景,则为目标设备的 ID;如果为本地设备场景,则不需要填写。

- path：资源的路径信息，代表特定资源的位置信息。
- query：查询参数。
- fragment：可用于指示要访问的子资源。

4.4.2　URI 示例

- 跨设备场景，dataability://device_id/com.domainname.dataability.persondata/person/10。
- 本地设备，dataability:///com.domainname.dataability.persondata/person/10。

其中，本地设备的 device_id 字段为空，因此在 dataability：后面有 3 个"/"。

4.4.3　创建 Data Ability

下面创建一个新的 Data Ability，在当前工程的主目录（entry＞src＞main＞java＞com.xxx.xxx）上右击，从弹出的快捷菜单中选择 New＞Ability＞Empty Data Ability，如图 4-33 所示。

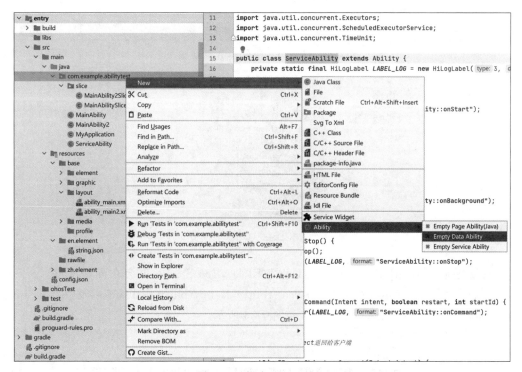

图 4-33　创建 Data Ability

然后，进入如图 4-34 所示的界面。

其中，Data Name 里填写需要自定义的 Data Ability 的名称，Package name 里填写新建的 Data Ability 需要放在哪个包下，如果填写的包名不存在，HUAWEI DevEco Studio 会自动创建该包。填写完成后单击 Finish 按钮，最后如图 4-35 所示。

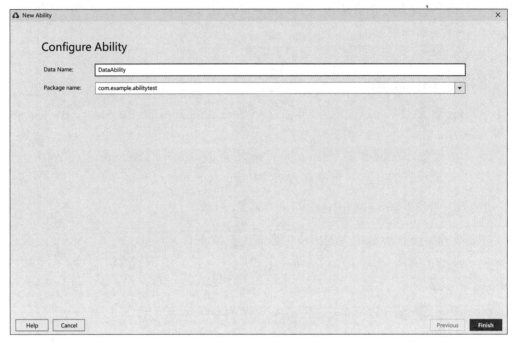

图 4-34 新建 Data Ability 的相关信息

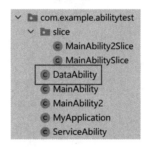

图 4-35 新建 Data Ability 的相关文件

在创建完 Data Ability 的同时，HUAWEI DevEco Studio 也会在配置文件（config.json）中注册该 Data Ability，相比 Service Ability，Data Ability 多了 uri 以及 permissions 属性，而且 type 的属性值是 data，具体如图 4-36 所示。

```
{
  "permissions": [
    "com.example.dataabilitytest.DataAbilityShellProvider.PROVIDER"
  ],
  "name": "com.example.dataabilitytest.DataAbility",
  "icon": "$media:icon",
  "description": "hap sample empty provider",
  "type": "data",
  "uri": "dataability://com.example.dataabilitytest.DataAbility"
}
```

图 4-36 新建 Data Ability 的配置信息

4.4.4　Data Ability 相关类

开发者可以通过 DataAbilityHelper 类访问当前应用或其他应用提供的共享数据。DataAbilityHelper 作为客户端，与提供方的 Data Ability 进行通信。Data Ability 接收到请求后，执行相应的处理，并返回结果。DataAbilityHelper 提供了一系列与 Data Ability 对应的方法。

DataAbilityHelper 为开发者提供了 creator() 方法来创建 DataAbilityHelper 实例。该方法为静态方法，有多个重载。最常见的方法是通过传入一个 Context 对象创建 DataAbilityHelper 对象，如程序清单 4-42 所示。

> 程序清单 4-42：hmos\ch04\01\AbilityTest\entry\src\main\java\com\example\abilitytest\DataAbility.java

```
DataAbilityHelper dataAbilityHelper = DataAbilityHelper.creator(this);
                        //用this指代当前上下文环境，即Context对象的实例
```

DataAbilityHelper 为开发者提供了增加、删除、修改、查询以及批量处理等方法来操作数据库。DataAbilityHelper 类定义的常用操作方法如表 4-1 所示。

表 4-1　DataAbilityHelper 类定义的常用操作方法

方　　法	描　　述
int insert(Uri uri, ValuesBucket value)	向数据库中插入单条数据
int batchInsert(Uri uri, ValuesBucket[] values)	向数据库中插入多条数据
int delete(Uri uri, DataAbilityPredicates predicates)	删除一条或多条数据
int update(Uri uri, ValuesBucket value, DataAbilityPredicates predicates)	更新数据库
DataAbilityResult[] executeBatch(ArrayList<DataAbilityOperation> operations)	批量操作数据库
void registerObserver(Uri uri, IDataAbilityObserver dataObserver)	根据 URL 指定订阅的数据表
void unregisterObserver(Uri uri, IDataAbilityObserver dataObserver)	根据 URL 指定取消订阅的数据表
staticDataAbilityHelper creator(Context context)	创建 DataAbilityHelper 实例
int batchInsert(Uri uri, ValuesBucket[] values)	批量插入方法，和 insert() 类似

开发者可以通过 DataAbilityPredicates 类构建删除、更新和查询的条件。例如，构造查询条件查询 userId 在 101～103 的数据，详细代码如程序清单 4-43 所示。

> 程序清单 4-43：hmos\ch04\01\AbilityTest\entry\src\main\java\com\example\abilitytest\DataAbility.java

```
1  // 构造查询条件
```

```
2    DataAbilityPredicates predicates = new DataAbilityPredicates();
3    predicates.between("userId", 101, 103);
```

DataAbilityPredicates 类定义的常用操作方法如表 4-2 所示。

表 4-2 DataAbilityPredicates 类定义的常用操作方法

方法	描述
DataAbilityPredicates between(String field, int low, int high)	设置谓词条件,满足 field 字段在最小值 low 和最大值 high 之间
DataAbilityPredicates notBetween(String field, int low, int high)	设置谓词条件,满足 field 字段不在最小值 low 和最大值 high 之间
DataAbilityPredicates contains(String field, String value)	设置谓词条件,满足 field 字段包含 value 值
DataAbilityPredicates equalTo(String field, int value)	设置谓词条件,满足 field 字段与 value 值相等
DataAbilityPredicates notEqualTo(String field, int value)	设置谓词条件,满足 field 字段与 value 值不相等
DataAbilityPredicates beginsWith(String field, String value)	设置谓词条件,满足 field 字段以 value 值开头
DataAbilityPredicates orderByAsc(String field)	设置谓词条件,根据 field 字段升序排列
DataAbilityPredicates orderByDesc(String field)	设置谓词条件,根据 field 字段降序排列
DataAbilityPredicates greaterThan(String field, String value)	设置谓词条件,满足 field 字段比 value 值大
DataAbilityPredicates lessThan(String field, String value)	设置谓词条件,满足 field 字段比 value 值小

在 Data Ability 中,需要插入的数据应由 ValuesBucket 封装,ValuesBucket 是以键-值对的形式存储的。程序清单 4-44 所示以批量插入数据为例来示范 ValuesBucket 的使用。

程序清单 4-44:hmos\ch04\01\AbilityTest\entry\src\main\java\com\example\abilitytest\DataAbility.java

```
1    //创建 DataAbilityHelper 对象
2    DataAbilityHelper dataAbilityHelper = DataAbilityHelper.creator(this);
3    // 构造插入数据
4    ValuesBucket[] valuesBucket = new ValuesBucket[2];
5    valuesBucket[0] = new ValuesBucket();
6    valuesBucket[0].putString("name", "Tom");
7    valuesBucket[0].putInteger("age", 12);
8    valuesBucket[1] = new ValuesBucket();
9    valuesBucket[1].putString("name", "Tom1");
```

```
10    valuesBucket[1].putInteger("age", 16);
11    //批量插入
12    dataAbilityHelper.batchInsert(uri, valuesBucket);
```

基于 Data Ability 对数据进行操作的具体操作可以参考 5.1.5 节——基于 Data Ability 的关系数据库操作案例。

4.5 本章小结

本章详细地介绍了 Ability 的相关知识，包括如何开发自己的 Ability、如何在 config.json 文件中配置不同模板的 Ability，以及 3 种不同模板的 Ability 的作用，介绍了 Page Ability 和 Service Ability 的生命周期、AbilitySlice 间导航和不同 Page Ability 间导航的方式、启动和停止本地或远程 Service Ability 的方式、启动前台 Service Ability 的方法，以及 Data Ability 的基本使用方式。

4.6 课后习题

1. Ability 的 2 种类型是_____和_____。
2. Ability 的 3 种模板是_____、_____和_____。
3. Page Ability 的 4 种状态是_____、_____、_____和_____。
4. 以下不属于 Page Ability 生命周期的回调方法的是(   )。
   A. onStart()   B. onActive()
   C. onBackground()   D. onFinish()
5. Page Ability 间的导航使用 startAbilityForResult()方法，获得返回结果的回调方法是(   )。
   A. onAbilityResult()   B. terminateAbility()
   C. goToAbility()   D. stopAbility()
6. 对于 Page Ability 中一些较为耗时的状态,保存操作最好在方法(   )中进行。
   A. onStart()   B. onActive()
   C. onBackground()   D. onFinish()
7. 以下选项中,(   )不是启动或停止 Service Ability 的方法。
   A. startAbility()   B. onResult()   C. onStart()   D. onInactive()
8. 简要描述 Page Ability、AbilitySlice 和 Service Ability 的生命周期。
9. 什么是前台 Ability Service？它与后台 Ability Service 的区别是什么？

第 5 章

数 据 管 理

课程思政 5

本章要点

- 关系数据库的使用
- 分布式数据服务

本章知识结构图（见图 5-1）

图 5-1　本章知识结构图

本章示例(见图 5-2)

图 5-2　本章示例图

一个较好的应用程序,应该能够为用户保存个性化的设置,能够保存用户的使用记录,而这些都离不开数据的存储,HarmonyOS 提供了关系数据库来存储信息。5.1.5 节将通过一个基于 Data Ability 的关系数据库操作案例详细讲解关系数据库的使用方式,读者可以根据该案例再次开发出其他更为复杂的应用程序。

HarmonyOS 还为开发者提供了分布式数据服务,当开发者想在两台或几台设备上共享同一数据时,就可以通过使用分布式数据服务来完成,5.2.4 节将通过一个分布式数据服务案例详细讲解 HarmonyOS 的分布式数据服务的使用方式。

5.1　关系数据库

5.1.1　关系数据库介绍

关系数据库(Relational Database,RDB)是指采用关系模型组织数据的数据库,是创建在关系模型基础上的数据库,以行和列的形式存储数据,方便用户理解。关系数据库可以理解为由二维表及其之间的关系组成的一个数据组织。

HarmonyOS 关系数据库基于 SQLite 组件实现,提供了一套完整的对本地数据库进行管理的机制,对外提供了一系列的增加、删除、修改、查询接口,也可以直接运行开发者输入的 SQL 语句来满足复杂的场景需要。

5.1.2 约束与限制

- 数据库中连接池的最大数量是 4 个，用以管理用户的读写操作。
- 为保证数据的准确性，数据库同一时间只能支持一个写操作。

5.1.3 关系数据库相关类

为了操作和管理关系数据库，HarmonyOS 为开发者提供了一些相关类，常用的有 DatabaseHelper、RdbOpenCallback、RdbStore、RdbPredicates、RawRdbPredicates、ResultSet。

DatabaseHelper 是 HarmonyOS 提供的管理数据库的工具类，主要用于关系数据库的创建、打开和删除。DatabaseHelper 类定义的常用操作方法如表 5-1 所示。

表 5-1 DatabaseHelper 类定义的常用操作方法

方　　法	描　　述
publicRdbStore getRdbStore(StoreConfig config，int version，RdbOpenCallback openCallback，ResultSetHook resultSetHook)	根据配置创建或打开数据库
public booleandeleteRdbStore(String name)	删除指定的数据库

RdbOpenCallback 是 HarmonyOS 为开发者提供的管理数据库创建、升级和降级的工具类。开发者可以创建一个子类来实现 onCreate() 和 onUpgrade() 等方法，如果一个数据库已经存在，它将被打开；如果不存在数据库，将创建一个新数据库。在数据库升级过程中，也会调用此类的方法。RdbOpenCallback 类定义的常用操作方法如表 5-2 所示。

表 5-2 RdbOpenCallback 类定义的常用操作方法

方　　法	描　　述
public abstract void onCreate(RdbStore store)	数据库创建时被调用，开发者可以在该方法中初始化表结构，并添加一些应用使用到的初始化数据
public abstract voidonUpgrade(RdbStore store，int currentVersion，int targetVersion)	数据库需要升级时被回调
void onOpen(RdbStore store)	打开数据库时调用

RdbStore 是 HarmonyOS 提供的用来对关系数据库进行增加、删除、修改、查询操作的工具类。RdbStore 类定义的常用操作方法如表 5-3 所示。

表 5-3 RdbStore 类定义的常用操作方法

方　　法	描　　述
long insert(String table，ValuesBucket initialValues)	向数据库插入数据
int update(ValuesBucket values，AbsRdbPredicates predicates)	更新数据库表中符合谓词指定条件的数据
ResultSet query(AbsRdbPredicates predicates，String[] columns)	查询数据

续表

方　　法	描　　述
ResultSet querySql(String sql, String[] sqlArgs)	执行原生的用于查询操作的 SQL 语句
int delete(AbsRdbPredicates predicates)	删除数据
void executeSql(String sqlString)	执行原生的 SQL 语句

AbsRdbPredicates 是关系数据库提供的用于设置数据库操作条件的谓词，其中包括两个实现子类 RdbPredicates 和 RawRdbPredicates。

- RdbPredicates：开发者无须编写复杂的 SQL 语句，仅通过调用该类中条件相关的方法，如 equalTo()、notEqualTo()、groupBy()、orderByAsc()、beginsWith()等，就可自动完成 SQL 语句拼接，方便用户聚焦业务操作。
- RawRdbPredicates：可满足复杂 SQL 语句的场景，支持开发者自己设置 where 条件子句和 whereArgs 参数，不支持 equalTo 等条件接口的使用。

RdbPredicates 类定义的常用操作方法如表 5-4 所示。RawRdbPredicates 类定义的常用操作方法如表 5-5 所示。

表 5-4　RdbPredicates 类定义的常用操作方法

方　　法	描　　述
RdbPredicates equalTo(String field, String value)	设置谓词条件，满足 field 字段与 value 值相等
RdbPredicates notEqualTo(String field, String value)	设置谓词条件，满足 field 字段与 value 值不相等
RdbPredicates beginsWith(String field, String value)	设置谓词条件，满足 field 字段以 value 值开头
RdbPredicates between(String field, int low, int high)	设置谓词条件，满足 field 字段在最小值 low 和最大值 high 之间
RdbPredicates notBetween(String field, int low, int high)	设置谓词条件，满足 field 字段不在最小值 low 和最大值 high 之间
RdbPredicates orderByAsc(String field)	设置谓词条件，根据 field 字段升序排列
RdbPredicates orderByDesc(String field)	设置谓词条件，根据 field 字段降序排列
RdbPredicates greaterThan(String field, String value)	设置谓词条件，满足 field 字段比 value 值大
RdbPredicates lessThan(String field, String value)	设置谓词条件，满足 field 字段比 value 值小

表 5-5　RawRdbPredicates 类定义的常用操作方法

方　　法	描　　述
void setWhereClause(String whereClause)	设置 where 条件子句
void setWhereArgs(List<String> whereArgs)	设置 whereArgs 参数，该值表示 where 子句中占位符的值

ResultSet 是关系数据库提供查询返回的结果集，它指向查询结果中的一行数据，供

用户对查询结果进行遍历和访问。ResultSet 类定义的常用操作方法如表 5-6 所示。

表 5-6 ResultSet 类定义的常用操作方法

方　　法	描　　述
boolean goTo(int offset)	从结果集当前位置移动指定偏移量
boolean goToRow(int position)	将结果集移动到指定位置
boolean goToFirstRow()	将结果集移动到第一行
boolean goToLastRow()	将结果集移动到最后一行
boolean goToNextRow()	将结果集向后移动一行
boolean goToPreviousRow()	将结果集向前移动一行
boolean isStarted()	判断结果集是否被移动过
boolean isEnded()	判断结果集当前位置是否在最后一行之后
boolean isAtFirstRow()	判断结果集当前位置是否在第一行
boolean isAtLastRow()	判断结果集当前位置是否在最后一行
int getRowCount()	获取当前结果集中的记录条数
int getColumnCount()	获取结果集中的列数
String getString(int columnIndex)	获取当前行指定列的值，以 String 类型返回
int getInt(int i)	获取当前行指定列的值，以 int 类型返回
void close()	关闭结果集
int getColumnIndexForName(String s)	获取结果集中列表名为 s 的列号
String getColumnNameForIndex(int i)	获取结果集中列号为 i 的列表名

5.1.4 关系数据库开发步骤

下面讲解关系数据库的创建，以及数据的增加、删除、修改、查询等基本步骤。

（1）首先创建数据库，配置数据库的相关信息，初始化数据库的表结构和相关数据，示例代码如程序清单 5-1 所示。

程序清单 5-1　hmos\ch05\01\RelationalDatabaseTest\entry\src\main\java\com\example\relationaldatabasetest\slice\MainAbilitySlice.java

```
1    StoreConfig storeConfig = StoreConfig.newDefaultConfig("RdbStoreTest.db");
2    DatabaseHelper databaseHelper= new DatabaseHelper(context);
3    private static RdbOpenCallback rdbOpenCallback = new RdbOpenCallback() {
4        @Override
5        public void onCreate(RdbStore rdbStore) {
6            //创建表
```

```
7          rdbStore.executeSql("CREATE TABLE IF NOT EXISTS testdb (userId
INTEGER PRIMARY KEY AUTOINCREMENT, userName TEXT NOT NULL, userAge
INTEGER)");
8       }
9       @Override
10      public void onUpgrade(RdbStore rdbStore, int oldVersion, int
newVersion) {
11          //升级数据库操作
12      }
13  };
14  RdbStore rdbStore = databaseHelper.getRdbStore(storeConfig, 1,
rdbOpenCallback, null);
```

（2）插入数据操作，以 ValuesBucket 形式构造要插入的数据，示例代码如程序清单 5-2 所示。

程序清单 5-2：hmos\ch05\01\RelationalDatabaseTest\entry\src\main\java\com\example\relationaldatabasetest\slice\MainAbilitySlice.java

```
1   ValuesBucket values = new ValuesBucket();
2   values.putInteger("userId", 1);
3   values.putString("userName", "Jerry");
4   values.putInteger("userAge", 18);
5   long id = rdbStore.insert("User", values);
6   //使用SQL语句插入数据
7   rdbStore.executeSql("insert into User (userId,userName,userAge) values
(2,'Tom',20)");
```

（3）删除数据操作，详细代码如程序清单 5-3 所示。

程序清单 5-3：hmos\ch05\01\RelationalDatabaseTest\entry\src\main\java\com\example\relationaldatabasetest\slice\MainAbilitySlice.java

```
1   RdbPredicates rdbPredicates = new RdbPredicates("User").equalTo
("userName","Jerry");
2   int i = rdbStore.delete(rdbPredicates);
```

（4）查询数据操作，详细代码如程序清单 5-4 所示。

程序清单 5-4：hmos\ch05\01\RelationalDatabaseTest\entry\src\main\java\com\example\relationaldatabasetest\slice\MainAbilitySlice.java

```
1   //按条件查询
2   String[] columns = new String[] {"userId", "userName", "userAge"};
```

```
3    RdPredicates rdbPredicates = new RdbPredicates("User").equalTo
     ("userAge", 18).orderByAsc("userId");
4    ResultSet resultSet = store.query(rdbPredicates, columns);
5    resultSet.goToNextRow();
6    //查询所有数据
7    String[] columns = new String[]{"userId","userName","userAge"};
8    RdbPredicates rdbPredicates = new RdbPredicates("User");//构建查询谓词
9    ResultSet resultSet = rdbCreateDb().query(rdbPredicates,columns);
10   while (resultSet.goToNextRow()){
11       int userId = resultSet.getInt(resultSet.getColumnIndexForName
     ("userId"));
12       String userName = resultSet.getString(resultSet.
     getColumnIndexForName("userName"));
13       int userAge = resultSet.getInt(resultSet.getColumnIndexForName
     ("userAge"));
14   }
```

(5) 更新数据操作,详细代码如程序清单 5-5 所示。

程序清单 5-5: hmos\ch05\01\RelationalDatabaseTest\entry\src\main\java\
com\example\relationaldatabasetest\slice\MainAbilitySlice.java

```
1    RdbPredicates rdbPredicates = new RdbPredicates("User").equalTo
     ("userName","Tom");
2    ValuesBucket values = new ValuesBucket();
3    values.putString("userName","Lee");
4    //更新数据
5    rdbStore.update(values,rdbPredicates);
```

5.1.5 基于 Data Ability 的关系数据库操作案例

前面讲解了操作 HarmonyOS 关系数据库的相关类以及关系数据库的基本开发步骤。本节将实现基于 Data Ability 创建数据库服务,对外提供访问数据库的服务端接口(PersonInfoDataAbility 类),并在 MainAbilitySlice 中通过 DataAbilityHelper 与提供方的 Data Ability 进行通信,最后通过日志查看数据库操作结果。

1. 创建一个 Data Ability

在新建的名为 RelationalDatabaseTest 的工程的主目录下添加一个名为 dataability 的包,在 dataability 下添加一个名为 PersonInfoDataAbility 的 Empty Data Ability,用于存储人员信息数据库并提供接口。HUAWEI DevEco Studio 将会自动生成数据库的 CRUD(增加、删除、修改、查询)方法,详细代码如程序清单 5-6 所示。

程序清单 5-6：hmos\ch05\01\RelationalDatabaseTest\entry\src\main\java\com\example\relationaldatabasetest\dataability\PersonInfoDataAbility.java

```java
1   public class PersonInfoDataAbility extends Ability {
2       private static final HiLogLabel LABEL_LOG = new HiLogLabel(3, 0xD001100, "Demo");
3
4       @Override
5       public void onStart(Intent intent) {
6           super.onStart(intent);
7           HiLog.info(LABEL_LOG, "PersonInfoDataAbility onStart");
8       }
9
10      @Override
11      public ResultSet query(Uri uri, String[] columns, DataAbilityPredicates predicates) {
12          return null;
13      }
14
15      @Override
16      public int insert(Uri uri, ValuesBucket value) {
17          HiLog.info(LABEL_LOG, "PersonInfoDataAbility insert");
18          return 999;
19      }
20
21      @Override
22      public int delete(Uri uri, DataAbilityPredicates predicates) {
23          return 0;
24      }
25
26      @Override
27      public int update(Uri uri, ValuesBucket value, DataAbilityPredicates predicates) {
28          return 0;
29      }
30  }
```

同时，HUAWEI DevEco Studio 将自动在工程配置文件 config.json 中添加如程序清单 5-7 所示的配置。

程序清单 5-7：hmos\ch05\01\RelationalDatabaseTest\entry\src\main\config.json

```
1   {
2       "permissions": [
```

```
3           "com.example.relationaldatabasetest.DataAbilityShellProvider.
      PROVIDER"
4       ],
5       "name": "com.example.relationaldatabasetest.dataability.
      PersonInfoDataAbility",
6       "icon": "$media:icon",
7       "description": "empty provider",
8       "type": "data",
9       "uri": "dataability://com.example.relationaldatabasetest.dataability.
      PersonInfoDataAbility"
10    }
```

2. 定义 Data Ability 数据库相关常量

为了方便后续操作,在 PersonInfoDataAbility 类中定义数据库的相关常量,包括数据库的库名、表名、表字段名、数据库版本号等,详细代码如程序清单 5-8 所示。

程序清单 5-8:hmos\ch05\01\RelationalDatabaseTest\entry\src\main\java\com\example\relationaldatabasetest\dataability\PersonInfoDataAbility.java

```
1   public class PersonInfoDataAbility extends Ability {
2   ...
3       private static final String DB_NAME = "personinfodataability.db";
                                                                //数据库名
4       private static final String DB_TAB_NAME = "personinfo";  //表名
5       private static final String DB_COLUMN_PERSON_ID = "id";
6       private static final String DB_COLUMN_NAME = "name";
7       private static final String DB_COLUMN_GENDER = "gender";
8       private static final String DB_COLUMN_AGE = "age";
9       private static final int DB_VERSION = 1;
10  ...
11  }
```

3. 创建关系数据库

在 PersonInfoDataAbility 类中定义 RdbStore 变量,并通过 RdbOpenCallback 创建数据库 personinfodataability.db 及表 personinfo,详细代码如程序清单 5-9 所示。

程序清单 5-9:hmos\ch05\01\RelationalDatabaseTest\entry\src\main\java\com\example\relationaldatabasetest\dataability\PersonInfoDataAbility.java

```
1   private StoreConfig config = StoreConfig.newDefaultConfig(DB_NAME);
2   private RdbStore rdbStore;
3   private RdbOpenCallback rdbOpenCallback = new RdbOpenCallback() {
4       @Override
5       public void onCreate(RdbStore store) {
```

```
6          store.executeSql("create table if not exists "
7                  + DB_TAB_NAME + " ("
8                  + DB_COLUMN_PERSON_ID + " integer primary key, "
9                  + DB_COLUMN_NAME + " text not null, "
10                 + DB_COLUMN_GENDER + " text not null, "
11                 + DB_COLUMN_AGE + " integer)");
12     }
13
14     @Override
15     public void onUpgrade(RdbStore store, int oldVersion, int
   newVersion) {
16     }
17 };
```

4. 初始化数据库连接

在程序应用启动时，系统会调用 Data Ability 中的 onStart() 方法创建 Data 实例。在此方法中，需要创建数据库连接，并获取连接对象，以便后续对数据库进行操作，详细代码如程序清单 5-10 所示。

程序清单 5-10：hmos\ch05\01\RelationalDatabaseTest\entry\src\main\java\com\example\relationaldatabasetest\dataability\PersonInfoDataAbility.java

```
1  @Override
2  public void onStart(Intent intent) {
3      super.onStart(intent);
4      HiLog.info(LABEL_LOG, "PersonInfoDataAbility onStart");
5      DatabaseHelper databaseHelper = new DatabaseHelper(this);
6      rdbStore = databaseHelper.getRdbStore(config, DB_VERSION,
   rdbOpenCallback, null);
7  }
```

5. 重写数据库操作方法

创建 Data Ability 时，HUAWEI DevEco Studio 自动生成数据库增加、删除、修改、查询的空方法，开发者可以按照自身需要重写相关方法。

(1) query()：查询数据方法。

该方法接收 3 个参数，分别为 uri（查询的目标路径）、columns（查询的列名）、predicates（查询条件）。其中查询条件由 DataAbilityPredicates 类构建，详细代码如程序清单 5-11 所示。

程序清单 5-11：hmos\ch05\01\RelationalDatabaseTest\entry\src\main\java\com\example\relationaldatabasetest\dataability\PersonInfoDataAbility.java

```
1  @Override
```

```
2   public ResultSet query(Uri uri, String[] columns, DataAbilityPredicates
    predicates) {
3       RdbPredicates rdbPredicates = DataAbilityUtils.createRdbPredicates
    (predicates, DB_TAB_NAME);
4       ResultSet resultSet = rdbStore.query(rdbPredicates, columns);
5       if (resultSet == null) {
6           HiLog.info(LABEL_LOG, "resultSet is null");
7       }
8       return resultSet;
9   }
```

(2) insert()：新增数据方法，该方法返回一个 int 类型的值用于标识结果。

该方法接收两个参数，分别为 uri（插入的目标路径）、value（插入的数据值）。其中，插入的数据由 ValuesBucket 封装，服务端可以从该参数中解析出对应的属性，然后插入数据库中，详细代码如程序清单 5-12 所示。

程序清单 5-12：hmos\ch05\01\RelationalDatabaseTest\entry\src\main\java\com\example\relationaldatabasetest\dataability\PersonInfoDataAbility.java

```
1   @Override
2   public int insert(Uri uri, ValuesBucket value) {
3       HiLog.info(LABEL_LOG, "PersonInfoDataAbility insert");
4       String path = uri.getLastPath();
5       if (!"person".equals(path)) {
6           HiLog.info(LABEL_LOG, "DataAbility insert path is not matched");
7           return -1;
8       }
9       ValuesBucket values = new ValuesBucket();
10      values.putInteger(DB_COLUMN_PERSON_ID, value.getInteger(DB_COLUMN_PERSON_ID));
11      values.putString(DB_COLUMN_NAME, value.getString(DB_COLUMN_NAME));
12      values.putString(DB_COLUMN_GENDER, value.getString(DB_COLUMN_GENDER));
13      values.putInteger(DB_COLUMN_AGE, value.getInteger(DB_COLUMN_AGE));
14      int index = (int) rdbStore.insert(DB_TAB_NAME, values);
15      DataAbilityHelper.creator(this, uri).notifyChange(uri);
16      return index;
17  }
```

(3) delete()：删除数据方法，该方法返回一个 int 类型的值用于标识结果。

该方法接收两个参数，分别为 uri（删除的目标路径）、predicates（删除条件）。删除条件由 DataAbilityPredicates 类构建，服务端在接收到该删除条件参数之后可以从中解析出要删除的数据，然后到数据库中执行，详细代码如程序清单 5-13 所示。

程序清单 5-13：hmos\ch05\01\RelationalDatabaseTest\entry\src\main\java\com\example\relationaldatabasetest\dataability\PersonInfoDataAbility.java

```
1   @Override
2   public int delete(Uri uri, DataAbilityPredicates predicates) {
3       RdbPredicates rdbPredicates = DataAbilityUtils.createRdbPredicates(predicates, DB_TAB_NAME);
4       int index = rdbStore.delete(rdbPredicates);
5       HiLog.info(LABEL_LOG, "delete: " + index);
6       DataAbilityHelper.creator(this, uri).notifyChange(uri);
7       return index;
8   }
```

(4) update()：更新数据方法。

该方法接收 3 个参数，分别是更新的目标路径、更新的数据值以及更新条件。开发者可以在 ValuesBucket 参数中指定要更新的数据，在 DataAbilityPredicates 中构建更新的条件等。此方法返回一个 int 类型的值用于标识结果。更新 person 表数据的详细代码如程序清单 5-14 所示。

程序清单 5-14：hmos\ch05\01\RelationalDatabaseTest\entry\src\main\java\com\example\relationaldatabasetest\dataability\PersonInfoDataAbility.java

```
1   @Override
2   public int update(Uri uri, ValuesBucket value, DataAbilityPredicates predicates) {
3       RdbPredicates rdbPredicates = DataAbilityUtils.createRdbPredicates(predicates, DB_TAB_NAME);
4       int index = rdbStore.update(value, rdbPredicates);
5       HiLog.info(LABEL_LOG, "update: " + index);
6       DataAbilityHelper.creator(this, uri).notifyChange(uri);
7       return index;
8   }
```

6．访问 Data Ability

开发者可以通过 DataAbilityHelper 类访问当前应用提供的共享数据。DataAbilityHelper 提供了一系列与 Data Ability 通信的方法。下面介绍 DataAbilityHelper 具体的使用步骤。

(1) 声明使用权限。

如果待访问的 Data Ability 声明了访问需要权限，则访问此 Data Ability 需要在工程配置文件 config.json 中声明需要此权限，如程序清单 5-15 所示。如果待访问的 Data Ability 由本应用创建，则可以不声明该权限。

程序清单 5-15：hmos\ch05\01\RelationalDatabaseTest\entry\src\main\config.json

```
1  "reqPermissions": [
2    {
3      "name": " com.example.relationaldatabasetest.
   DataAbilityShellProvider.PROVIDER "
4    }
5  ]
```

（2）创建 DataAbilityHelper。

DataAbilityHelper 为开发者提供了 creator()方法来创建 DataAbilityHelper 实例，该方法为静态方法，有多个重载。最常见的方法是：通过传入一个 context 对象创建 DataAbilityHelper 对象。为了方便操作，直接在 MainAbilitySlice 中定义 DataAbilityHelper 变量，并在 onStart()方法中创建其实例，详细代码如程序清单 5-16 所示。

程序清单 5-16：hmos\ch05\01\RelationalDatabaseTest\entry\src\main\java\com\example\relationaldatabasetest\slice\MainAbilitySlice.java

```
1  public class MainAbilitySlice extends AbilitySlice {
2      private static final HiLogLabel LABEL_LOG = new HiLogLabel(3,
   0xD001100, "MainAbilitySlice");
3      ...
4      private DataAbilityHelper databaseHelper;
5
6      @Override
7      public void onStart(Intent intent) {
8          super.onStart(intent);
9          super.setUIContent(ResourceTable.Layout_ability_main);
10         databaseHelper = DataAbilityHelper.creator(this);
11     }
12     ...
13 }
```

（3）访问 DataAbility。

DataAbilityHelper 为开发者提供了增加、删除、修改、查询以及批量处理等方法来操作数据库。同样，为了方便操作，直接在 MainAbilitySlice 中定义相关的常量及数据库的增加、删除、修改、查询方法，并通过系统提供的日志接口 HiLog.info 记录相关结果。

① 首先需要定义访问数据库的常量，详细代码如程序清单 5-17 所示。

程序清单 5-17：hmos\ch05\01\RelationalDatabaseTest\entry\src\main\java\com\example\relationaldatabasetest\slice\MainAbilitySlice.java

```
1  private static final String BASE_URI = "dataability:///com.example.
   relationaldatabasetest.dataability.PersonInfoDataAbility";
```

```
2    private static final String DATA_PATH = "/person";
3    private static final String DB_COLUMN_PERSON_ID = "id";
4    private static final String DB_COLUMN_NAME = "name";
5    private static final String DB_COLUMN_GENDER = "gender";
6    private static final String DB_COLUMN_AGE = "age";
```

② 编写查询数据的 query()方法,其中 uri 为目标资源路径,columns 为要查询的字段,开发者的查询条件可以通过 DataAbilityPredicates 构建。开发者无须编写复杂的 SQL 语句,仅通过调用该类中条件相关的方法就可自动完成 SQL 语句拼接。根据年龄区间、名字及性别等查询条件查询数据库的详细代码,如程序清单 5-18 所示。

程序清单 5-18: hmos\ch05\01\RelationalDatabaseTest\entry\src\main\java\com\example\relationaldatabasetest\slice\MainAbilitySlice.java

```
1    private void query() {
2        String[] columns = new String[] {DB_COLUMN_PERSON_ID,
3            DB_COLUMN_NAME, DB_COLUMN_GENDER, DB_COLUMN_AGE};
4        // 构造查询条件,设置年龄范围为 15~30
5        DataAbilityPredicates predicates = new DataAbilityPredicates();
6        predicates.between(DB_COLUMN_AGE, 15, 30);
7        try {
8            ResultSet resultSet = databaseHelper.query(Uri.parse(BASE_URI + DATA_PATH),
9                columns, predicates);
10           if (resultSet == null || resultSet.getRowCount() == 0) {
11               HiLog.info(LABEL_LOG, "query: resultSet is null or no result found");
12               return;
13           }
14           resultSet.goToFirstRow();
15           do {
16               int id = resultSet.getInt(resultSet.getColumnIndexForName(DB_COLUMN_PERSON_ID));
17               String name = resultSet.getString(resultSet.getColumnIndexForName(DB_COLUMN_NAME));
18               String gender = resultSet.getString(resultSet.getColumnIndexForName(DB_COLUMN_GENDER));
19               int age = resultSet.getInt(resultSet.getColumnIndexForName(DB_COLUMN_AGE));
20               HiLog.info(LABEL_LOG, "query: Id :" + id + " Name :" + name + " Gender :" + gender + " Age :" + age);
21           } while (resultSet.goToNextRow());
22       } catch (DataAbilityRemoteException | IllegalStateException exception) {
```

```
23            HiLog.error(LABEL_LOG, "query: dataRemote exception |
   illegalStateException");
24        }
25 }
```

③ 编写新增数据的 insert()方法,其中 uri 为目标资源路径,ValuesBucket 为要新增的对象。插入一条人员信息的详细代码如程序清单 5-19 所示。

程序清单 5-19 hmos\ch05\01\RelationalDatabaseTest\entry\src\main\java\com\example\relationaldatabasetest\slice\MainAbilitySlice.java

```
1  private void insert(int id, String name, String gender, int age) {
2      ValuesBucket valuesBucket = new ValuesBucket();
3      valuesBucket.putInteger(DB_COLUMN_PERSON_ID, id);
4      valuesBucket.putString(DB_COLUMN_NAME, name);
5      valuesBucket.putString(DB_COLUMN_GENDER, gender);
6      valuesBucket.putInteger(DB_COLUMN_AGE, age);
7      try {
8          if (databaseHelper.insert(Uri.parse(BASE_URI + DATA_PATH),
   valuesBucket) != -1) {
9              HiLog.info(LABEL_LOG, "insert successful");
10         }
11     } catch (DataAbilityRemoteException | IllegalStateException
   exception) {
12         HiLog.error(LABEL_LOG, "insert: dataRemote exception|
   illegalStateException");
13     }
14 }
```

④ 编写更新数据的 update()方法,更新数据通过 ValuesBucket 封装,更新条件由 DataAbilityPredicates 构建。例如,将 ID 为 103 的表格数据,姓名修改为"Jerry",年龄修改为 27,详细代码如程序清单 5-20 所示。

程序清单 5-20 hmos\ch05\01\RelationalDatabaseTest\entry\src\main\java\com\example\relationaldatabasetest\slice\MainAbilitySlice.java

```
1  private void update() {
2      DataAbilityPredicates predicates = new DataAbilityPredicates();
3      predicates.equalTo(DB_COLUMN_PERSON_ID, 103);
4      ValuesBucket valuesBucket = new ValuesBucket();
5      valuesBucket.putString(DB_COLUMN_NAME, "Jerry");
6      valuesBucket.putInteger(DB_COLUMN_AGE, 27);
7      try {
8          if (databaseHelper.update(Uri.parse(BASE_URI + DATA_PATH),
   valuesBucket, predicates) != -1) {
```

```
9              HiLog.info(LABEL_LOG, "update successful");
10         }
11     } catch (DataAbilityRemoteException | IllegalStateException
   exception) {
12         HiLog.error(LABEL_LOG, "update: dataRemote exception |
   illegalStateException");
13     }
14 }
```

⑤ 编写删除数据的 delete() 方法,删除条件由 DataAbilityPredicates 构建,其中 uri 为目标资源路径,predicates 为删除条件,详细代码如程序清单 5-21 所示。

程序清单 5-21:hmos\ch05\01\RelationalDatabaseTest\entry\src\main\java\com\example\relationaldatabasetest\slice\MainAbilitySlice.java

```
1  private void delete(int id) {
2      DataAbilityPredicates predicates = new DataAbilityPredicates()
3              .equalTo(DB_COLUMN_PERSON_ID, id);
4      try {
5          if (databaseHelper.delete(Uri.parse(BASE_URI + DATA_PATH),
   predicates) != -1) {
6              HiLog.info(LABEL_LOG, "delete successful");
7          }
8      } catch (DataAbilityRemoteException | IllegalStateException
   exception) {
9          HiLog.error(LABEL_LOG, "delete: dataRemote exception |
   illegalStateException");
10     }
11 }
```

7. 测试与运行效果

为了快速查看代码效果,在 MainAbilitySlice 的 onStart() 中调用定义好的 query()、insert()、update()、delete() 等方法,详细代码如程序清单 5-22 所示。

程序清单 5-22:hmos\ch05\01\RelationalDatabaseTest\entry\src\main\java\com\example\relationaldatabasetest\slice\MainAbilitySlice.java

```
1  @Override
2  public void onStart(Intent intent) {
3      super.onStart(intent);
4      super.setUIContent(ResourceTable.Layout_ability_main);
5      databaseHelper = DataAbilityHelper.creator(this);
6      query();
7      insert(101, "Alan", "male", 20);
```

```
8      insert(102, "Lily", "female", 21);
9      insert(103, "Lee", "male", 23);
10     query();                              //查看插入后的结果
11     update();
12     query();                              //查看更新后的结果
13     delete(101);                          //删除 ID 为 101 的数据
14     query();                              //查看删除后的结果
15   }
```

连接手机或者模拟器,运行该工程,通过 HUAWEI DevEco Studio HiLog 工具查看结果,显示结果如图 5-3 所示。

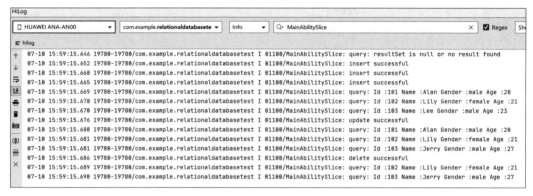

图 5-3　基于 Data Ability 的关系数据库操作案例

8. 订阅数据变化

通常情况下,当数据库表格的内容产生变化时,需要主动通知与该表格数据相关联的进程或者应用,从而使得相关进程或者应用接收到数据变化后完成相应的处理。对于数据提供方(PersonInfoDataAbility),按照开发者自身需求,可在它的 insert()、delete()、update()方法中加入 DataAbilityHelper.creator(this,uri).notifyChange(uri);这一语句,当数据库表格内容变化时,会通知数据订阅者。

对于数据接收方(MainAbilitySlice),可以通过 DataAbilityHelper 类提供的 IDataAbilityObserver()方法注册一个数据订阅者,并重写 onChange()方法。订阅者接收数据库变化的详细代码如程序清单 5-23 所示。

程序清单 5-23：hmos\ch05\01\RelationalDatabaseTest\entry\src\main\java\com\example\relationaldatabasetest\slice\MainAbilitySlice.java

```
1   IDataAbilityObserver dataAbilityObserver;
2   ...
3   private void personInfoDatabaseObserver() {
4       dataAbilityObserver = new IDataAbilityObserver() {
5           @Override
```

```
6         public void onChange() {
7             //订阅者接收目标数据表格产生变化的通知,通过查询获取最新的数据
8             HiLog.debug(LABEL_LOG, "数据更新!");
9         }
10     };
11     //根据 URL 指定订阅的数据表
12     databaseHelper.registerObserver(Uri.parse(BASE_URI + DATA_PATH),
   dataAbilityObserver);
13 }
```

当数据订阅者不需要在数据库表格内容变化时接收通知,可调用 unregisterObserver(Uri uri,IDataAbilityObserver dataObserver)方法取消订阅。

9. 示例的完整代码

基于 Data Ability 的关系数据库操作案例的主要文件 MainAbilitySlice.java 和 PersonInfoDataAbility.java 的完整代码如程序清单 5-24 和程序清单 5-25 所示。

程序清单 5-24：hmos\ch05\01\RelationalDatabaseTest\entry\src\main\java\com\example\relationaldatabasetest\slice\MainAbilitySlice.java

```
1  public class MainAbilitySlice extends AbilitySlice {
2      private static final HiLogLabel LABEL_LOG = new HiLogLabel(3,
   0xD001100, "MainAbilitySlice");
3      private DataAbilityHelper databaseHelper;
4      private static final String BASE_URI = "dataability:///com.example.
   relationaldatabasetest.dataability.PersonInfoDataAbility";
5      private static final String DATA_PATH = "/person";
6      private static final String DB_COLUMN_PERSON_ID = "id";
7      private static final String DB_COLUMN_NAME = "name";
8      private static final String DB_COLUMN_GENDER = "gender";
9      private static final String DB_COLUMN_AGE = "age";
10     IDataAbilityObserver dataAbilityObserver;
11
12     @Override
13     public void onStart(Intent intent) {
14         super.onStart(intent);
15         super.setUIContent(ResourceTable.Layout_ability_main);
16         databaseHelper = DataAbilityHelper.creator(this);
17         personInfoDatabaseObserver();
18         query();
19         insert(101, "Alan", "male", 20);
20         insert(102, "Lily", "female", 21);
21         insert(103, "Lee", "male", 23);
22         query();                            //查看插入后的结果
```

```
23          update();
24          query();                        //查询更新后的结果
25          delete(101);
26          query();                        //查看删除后的结果
27      }
28
29      //订阅数据变化
30      private void personInfoDatabaseObserver() {
31          dataAbilityObserver = new IDataAbilityObserver() {
32              @Override
33              public void onChange() {
34                  //订阅者接收目标数据表格产生变化的通知
35                  HiLog.debug(LABEL_LOG, "数据更新!");
36              }
37          };
38          //根据 uri 指定订阅的数据表
39          databaseHelper.registerObserver(Uri.parse(BASE_URI + DATA_PATH), dataAbilityObserver);
40      }
41
42      private void delete(int id) {
43          DataAbilityPredicates predicates = new DataAbilityPredicates()
44                  .equalTo(DB_COLUMN_PERSON_ID, id);
45          try {
46              if (databaseHelper.delete(Uri.parse(BASE_URI + DATA_PATH), predicates) != -1) {
47                  HiLog.info(LABEL_LOG, "delete successful");
48              }
49          } catch (DataAbilityRemoteException | IllegalStateException exception) {
50              HiLog.error(LABEL_LOG, "delete: dataRemote exception | illegalStateException");
51          }
52      }
53
54      private void update() {
55          DataAbilityPredicates predicates = new DataAbilityPredicates();
56          predicates.equalTo(DB_COLUMN_PERSON_ID, 103);
57          ValuesBucket valuesBucket = new ValuesBucket();
58          valuesBucket.putString(DB_COLUMN_NAME, "Jerry");
59          valuesBucket.putInteger(DB_COLUMN_AGE, 27);
60          try {
61              if (databaseHelper.update(Uri.parse(BASE_URI + DATA_PATH), valuesBucket, predicates) != -1) {
```

```
62              HiLog.info(LABEL_LOG, "update successful");
63          }
64      } catch (DataAbilityRemoteException | IllegalStateException exception) {
65          HiLog.error(LABEL_LOG, "update: dataRemote exception | illegalStateException");
66      }
67  }
68
69  private void insert(int id, String name, String gender, int age) {
70      ValuesBucket valuesBucket = new ValuesBucket();
71      valuesBucket.putInteger(DB_COLUMN_PERSON_ID, id);
72      valuesBucket.putString(DB_COLUMN_NAME, name);
73      valuesBucket.putString(DB_COLUMN_GENDER, gender);
74      valuesBucket.putInteger(DB_COLUMN_AGE, age);
75      try {
76          if (databaseHelper.insert(Uri.parse(BASE_URI + DATA_PATH), valuesBucket) != -1) {
77              HiLog.info(LABEL_LOG, "insert successful");
78          }
79      } catch (DataAbilityRemoteException | IllegalStateException exception) {
80          HiLog.error(LABEL_LOG, "insert: dataRemote exception| illegalStateException");
81          HiLog.error(LABEL_LOG, exception.toString());
82      }
83  }
84
85  private void query() {
86      String[] columns = new String[]{DB_COLUMN_PERSON_ID,
87          DB_COLUMN_NAME, DB_COLUMN_GENDER, DB_COLUMN_AGE};
88      //构造查询条件,设置年龄范围为15~30
89      DataAbilityPredicates predicates = new DataAbilityPredicates();
90      predicates.between(DB_COLUMN_AGE, 15, 30);
91      try {
92          ResultSet resultSet = databaseHelper.query(Uri.parse(BASE_URI + DATA_PATH),
93              columns, predicates);
94          if (resultSet == null || resultSet.getRowCount() == 0) {
95              HiLog.info(LABEL_LOG, "query: resultSet is null or no result found");
96              return;
97          }
```

```
 98                 resultSet.goToFirstRow();
 99                 do {
100                     int id = resultSet.getInt(resultSet.
    getColumnIndexForName(DB_COLUMN_PERSON_ID));
101                     String name = resultSet.getString(resultSet.
    getColumnIndexForName(DB_COLUMN_NAME));
102                     String gender = resultSet.getString(resultSet.
    getColumnIndexForName(DB_COLUMN_GENDER));
103                     int age = resultSet.getInt(resultSet.
    getColumnIndexForName(DB_COLUMN_AGE));
104                     HiLog.info(LABEL_LOG, "query: Id :" + id + " Name :" + name + "
    Gender :" + gender + " Age :" + age);
105                 } while (resultSet.goToNextRow());
106             } catch (DataAbilityRemoteException | IllegalStateException
    exception) {
107                 HiLog.error(LABEL_LOG, "query: dataRemote exception |
    illegalStateException");
108             }
109         }
110
111     @Override
112     public void onActive() {
113         super.onActive();
114     }
115
116     @Override
117     public void onForeground(Intent intent) {
118         super.onForeground(intent);
119     }
120 }
```

程序清单 5-28：hmos\ch05\01\RelationalDatabaseTest\entry\src\main\java\com\example\relationaldatabasetest\dataability\PersonInfoDataAbility.java

```
1  public class PersonInfoDataAbility extends Ability {
2      private static final HiLogLabel LABEL_LOG = new HiLogLabel(3,
    0xD001100, "Demo");
3      private static final String DB_NAME = "personinfodataability.db";
                                                                    //数据库名
4      private static final String DB_TAB_NAME = "personinfo";   //表名
5      private static final String DB_COLUMN_PERSON_ID = "id";
6      private static final String DB_COLUMN_NAME = "name";
7      private static final String DB_COLUMN_GENDER = "gender";
```

```
8       private static final String DB_COLUMN_AGE = "age";
9       private static final int DB_VERSION = 1;
10      private StoreConfig config = StoreConfig.newDefaultConfig(DB_NAME);
11      private RdbStore rdbStore;
12      private RdbOpenCallback rdbOpenCallback = new RdbOpenCallback() {
13
14          @Override
15          public void onCreate(RdbStore rdbStore) {
16              rdbStore.executeSql("create table if not exists "
17                      + DB_TAB_NAME + " ("
18                      + DB_COLUMN_PERSON_ID + " integer primary key, "
19                      + DB_COLUMN_NAME + " text not null, "
20                      + DB_COLUMN_GENDER + " text not null, "
21                      + DB_COLUMN_AGE + " integer)");
22          }
23
24          @Override
25          public void onUpgrade(RdbStore rdbStore, int i, int i1) {
26
27          }
28      };
29
30      @Override
31      public void onStart(Intent intent) {
32          super.onStart(intent);
33          HiLog.info(LABEL_LOG, "PersonInfoDataAbility onStart");
34          DatabaseHelper databaseHelper = new DatabaseHelper(this);
35          rdbStore = databaseHelper.getRdbStore(config, DB_VERSION, rdbOpenCallback, null);
36      }
37
38      @Override
39      public ResultSet query(Uri uri, String[] columns, DataAbilityPredicates predicates) {
40          RdbPredicates rdbPredicates=DataAbilityUtils.createRdbPredicates(predicates, DB_TAB_NAME);
41          ResultSet resultSet = rdbStore.query(rdbPredicates, columns);
42          if (resultSet == null) {
43              HiLog.info(LABEL_LOG, "resultSet is null");
44          }
45          return resultSet;
46      }
47
```

```java
48      @Override
49      public int insert(Uri uri, ValuesBucket value) {
50          HiLog.info(LABEL_LOG, "PersonInfoDataAbility insert");
51          String path = uri.getLastPath();
52          if (!"person".equals(path)) {
53              HiLog.info(LABEL_LOG, "DataAbility insert path is not matched");
54              return -1;
55          }
56          ValuesBucket values = new ValuesBucket();
57          values.putInteger(DB_COLUMN_PERSON_ID, value.getInteger(DB_COLUMN_PERSON_ID));
58          values.putString(DB_COLUMN_NAME, value.getString(DB_COLUMN_NAME));
59          values.putString(DB_COLUMN_GENDER, value.getString(DB_COLUMN_GENDER));
60          values.putInteger(DB_COLUMN_AGE, value.getInteger(DB_COLUMN_AGE));
61          int index = (int) rdbStore.insert(DB_TAB_NAME, values);
62          DataAbilityHelper.creator(this, uri).notifyChange(uri);
63          return index;
64      }
65
66      @Override
67      public int delete(Uri uri, DataAbilityPredicates predicates) {
68          RdbPredicates rdbPredicates=DataAbilityUtils.createRdbPredicates(predicates, DB_TAB_NAME);
69          int index = rdbStore.delete(rdbPredicates);
70          HiLog.info(LABEL_LOG, "delete: " + index);
71          DataAbilityHelper.creator(this, uri).notifyChange(uri);
72          return index;
73      }
74
75      @Override
76      public int update(Uri uri, ValuesBucket value, DataAbilityPredicates predicates) {
77          RdbPredicates rdbPredicates=DataAbilityUtils.createRdbPredicates(predicates, DB_TAB_NAME);
78          int index = rdbStore.update(value, rdbPredicates);
79          HiLog.info(LABEL_LOG, "update: " + index);
80          DataAbilityHelper.creator(this, uri).notifyChange(uri);
81          return index;
82      }
83
```

```
84        @Override
85        public FileDescriptor openFile(Uri uri, String mode) {
86            return null;
87        }
88
89        @Override
90        public String[] getFileTypes(Uri uri, String mimeTypeFilter) {
91            return new String[0];
92        }
93
94        @Override
95        public PacMap call(String method, String arg, PacMap extras) {
96            return null;
97        }
98
99        @Override
100       public String getType(Uri uri) {
101           return null;
102       }
103   }
```

5.2 分布式数据服务

5.2.1 分布式数据服务介绍

HarmonyOS 的分布式数据服务(Distributed Data Service,DDS)为应用程序提供不同设备间数据库数据分布式的能力。当开发者想在两台或几台设备上共享同一数据时,开发者可以通过调用分布式数据接口将数据保存到分布式数据库中来达到目的。为了保证不同应用之间的数据不能通过分布式数据服务互相访问,HarmonyOS 分布式数据服务会对属于不同应用的数据进行隔离。分布式数据服务虽然具有一定的存储功能,但不要把分布式数据库的存储能力当作本地数据库来存储数据。当开发者需要使用分布式数据服务完整功能时,需要在配置文件中申请 ohos.permission.DISTRIBUTED_DATASYNC 权限,且分布式数据服务的数据模型仅支持 KV 数据模型。

5.2.2 单版本分布式数据库

单版本分布式数据库是 HarmonyOS 提供的分布式数据库之一。单版本是指数据以单个 KV(Key-Value 数据模型)条目为单位的方式在本地保存,对每个 Key 最多只保存一个 Value 值,即 Key 值不能重复。当数据在本地被用户修改时,不管它是否已经被同步出去,均直接在这个条目上进行修改,同步时也是将当前最新一次修改的条目逐条同步至远端设备。

5.2.3 分布式数据服务相关类

为了操作和管理分布式数据库，HarmonyOS 为开发者提供了一些相关类，常用的有 KvManager、SingleKvStore、Options、KvManagerConfig、KvStoreObserver。

KvManager 是 HarmonyOS 提供的用来管理分布式数据库的工具类，主要用于分布式数据库的创建、关闭和删除。KvManager 类定义的常用操作方法如表 5-7 所示。

表 5-7　KvManager 类定义的常用操作方法

方　　法	描　　述
void closeKvStore(KvStore kvStore)	关闭分布式数据库
void deleteKvStore(String storeId)	删除分布式数据库
<KVSTORE extendsKvStore>KVSTORE getKvStore(Options options，String storeId)	根据 Options 配置创建和打开标识符为 storeId 的分布式数据库
List<String> getAllKvStoreId()	获取所有分布式数据库的 ID，这些数据库是使用 getKvStore() 方法创建的，而且没有通过调用 deleteKvStore() 方法删除

SingleKvStore 是 HarmonyOS 为开发者提供的与单版本分布式数据库相关的方法，主要用于创建和管理单版本分布式数据库。开发者可以通过使用 KvManager.getKvStore(Options options，String storeId) 方法创建一个单版本分布式数据库。单版本分布式数据库可以按时间顺序将数据同步到其他数据库，SingleKvStore 类定义的常用操作方法如表 5-8 所示。

表 5-8　SingleKvStore 类定义的常用操作方法

方　　法	描　　述
putBoolean(String key，boolean value) putInt(String key，int value) putFloat(String key，float value) putDouble(String key，double value) putString(String key，String value) putByteArray(String key，byte[] value) putBatch(List<Entry> entries)	插入和更新数据
delete(String key) deleteBatch(List<String> keys)	删除数据
getInt(String key) getFloat(String key) getDouble(String key) getString(String key) getByteArray(String key) getEntries(String keyPrefix)	查询数据

续表

方法	描述
sync(List<String>deviceIdList, SyncMode mode)	在手动模式下，触发数据库同步到指定的设备
sync(List<String>deviceIdList, SyncMode mode, int allowedDelayMs)	在手动模式下，延迟 allowedDelayMs 毫秒后将数据库同步到指定的设备
setSyncParam(int defaultAllowedDelayMs)	设置数据库同步的默认延迟
subscribe(SubscribeType subscribeType, KvStoreObserver observer)	订阅数据库中数据的变化
registerSyncCallback(SyncCallback syncCallback)	注册单版本分布式数据的同步回调
unRegisterSyncCallback()	注销单版本分布式数据的同步回调

Options 是 HarmonyOS 提供的用于创建分布式数据库的配置选项。如果缺少分布式数据库，可以设置是否创建另一个数据库、是否加密该数据库，以及数据库类型等。Options 类定义的常用操作方法如表 5-9 所示。

表 5-9 Options 类定义的常用操作方法

方法	描述
setCreateIfMissing(boolean isCreateIfMissing)	设置数据库不存在时是否创建
isCreateIfMissing()	检查数据库不存在时是否创建
setEncrypt(boolean isEncrypt)	设置数据库是否加密
isEncrypt()	检查数据库是否加密
setStoreType(KvStoreType storeType)	设置分布式数据库的类型
getStoreType()	获取分布式数据库的类型

KvManagerConfig 是为 KvManager 实例提供配置信息的类，通过传入 Context 参数来实例化。KvStoreObserver 是 SingleKvStore.subscribe() 方法的输入参数，该方法用于订阅分布式数据库，当分布式数据库中的数据更改时调用此方法。

5.2.4 单版本分布式数据服务案例

前面讲解了操作 HarmonyOS 分布式数据服务的相关类。本节将通过分布式数据服务实现自定义的数据在两台设备间的同步传输。

1. 设置布局文件

新建名为 DistributedDataServiceTest 的项目后，编写它的 ability_main.xml 文件，在布局文件中，设置两个输入框用于输入需要保存的 key 值和 value 值，再设置 3 个按钮，分别用来保存输入的数据、删除指定的数据和同步数据到其他设备，详细代码如程序清单 5-26 所示。

程序清单 5-26：hmos\ch05\02\DistributedDataServiceTest\entry\src\main\resources\base\layout\ability_main.xml

```xml
1    <?xml version="1.0" encoding="utf-8"?>
2    <DirectionalLayout
3        xmlns:ohos="http://schemas.huawei.com/res/ohos"
4        ohos:height="match_parent"
5        ohos:width="match_parent"
6        ohos:alignment="horizontal_center"
7        ohos:orientation="vertical">
8        <DirectionalLayout
9            ohos:height="match_content"
10           ohos:width="match_parent"
11           ohos:alignment="center"
12           ohos:orientation="horizontal"
13           ohos:padding="20vp">
14           <Text
15               ohos:height="match_content"
16               ohos:width="match_parent"
17               ohos:text="Key"
18               ohos:text_size="30vp"
19               ohos:weight="1"/>
20           <TextField
21               ohos:id="$+id:key_input"
22               ohos:height="match_content"
23               ohos:width="match_parent"
24               ohos:hint="key..."
25               ohos:left_padding="20vp"
26               ohos:text_alignment="vertical_center"
27               ohos:text_size="30fp"
28               ohos:weight="3"/>
29       </DirectionalLayout>
30       <DirectionalLayout
31           ohos:height="match_content"
32           ohos:width="match_parent"
33           ohos:alignment="center"
34           ohos:orientation="horizontal"
35           ohos:padding="20vp">
36           <Text
37               ohos:height="match_content"
38               ohos:width="match_parent"
39               ohos:text="Value"
40               ohos:text_size="30vp"
```

```
41              ohos:weight="1"/>
42          <TextField
43              ohos:id="$+id:value_input"
44              ohos:height="match_content"
45              ohos:width="match_parent"
46              ohos:hint="value..."
47              ohos:left_padding="20vp"
48              ohos:text_alignment="vertical_center"
49              ohos:text_size="30fp"
50              ohos:weight="3"/>
51      </DirectionalLayout>
52      <DirectionalLayout
53          ohos:height="match_content"
54          ohos:width="match_parent"
55          ohos:alignment="horizontal_center"
56          ohos:orientation="horizontal"
57          ohos:padding="20vp">
58          <Button
59              ohos:id="$+id:input_button"
60              ohos:height="match_content"
61              ohos:width="match_parent"
62              ohos:background_element="#fff000"
63              ohos:margin="10vp"
64              ohos:text="写入"
65              ohos:text_size="20vp"
66              ohos:weight="1"/>
67          <Button
68              ohos:id="$+id:delete_button"
69              ohos:height="match_content"
70              ohos:width="match_parent"
71              ohos:background_element="#fff000"
72              ohos:margin="10vp"
73              ohos:text="删除"
74              ohos:text_size="20vp"
75              ohos:weight="1"/>
76          <Button
77              ohos:id="$+id:sync_button"
78              ohos:height="match_content"
79              ohos:width="match_parent"
80              ohos:background_element="#fff000"
81              ohos:margin="10vp"
82              ohos:text="同步"
83              ohos:text_size="20vp"
```

```
84            ohos:weight="1"/>
85      </DirectionalLayout>
86      <Text
87          ohos:id="$+id:text_output"
88          ohos:height="match_content"
89          ohos:width="match_parent"
90          ohos:margin="10vp"
91          ohos:text_alignment="horizontal_center"
92          ohos:text_size="20vp"/>
93  </DirectionalLayout>
```

2. 实现分布式数据库需要申请的权限

为了实现分布式数据库，需要在"entry＞src＞main＞config.json"中申请 ohos.permission.DISTRIBUTED_DATASYNC 权限，以便允许不同设备间的数据交换，详细代码如程序清单 5-27 所示。

程序清单 5-27：hmos\ch05\02\DistributedDataServiceTest\entry\src\main\config.json

```
1   module": {
2   ...
3     "reqPermissions": [
4       {
5          "name": "ohos.permission.DISTRIBUTED_DATASYNC"
6       }
7     ]
8   }
```

3. 创建分布式数据库

在 MainAbilitySlice 中定义创建分布式数据库管理器实例 KvManager 的方法，详细代码如程序清单 5-28 所示。

程序清单 5-28：hmos\ch05\02\DistributedDataServiceTest\entry\src\main\
java\com\example\distributeddataservicetest\slice\MainAbilitySlice.java

```
1   private KvManager createManager() {
2       KvManager kvManager = null;
3       try {
4           KvManagerConfig kvManagerConfig = new KvManagerConfig(this);
5           kvManager = KvManagerFactory.getInstance().createKvManager(kvManagerConfig);
6       } catch (KvStoreException exception) {
7       }
8       return kvManager;
9   }
```

创建成功后,借助 KvManager 创建 SINGLE_VERSION 分布式数据库,详细代码如程序清单 5-29 所示。

程序清单 5-29: hmos\ch05\02\DistributedDataServiceTest\entry\src\main\java\com\example\distributeddataservicetest\slice\MainAbilitySlice.java

```java
1   private SingleKvStore createDb(KvManager kvManager) {
2       SingleKvStore singleKvStore = null;
3       try {
4           Options CREATE = new Options();
5           CREATE.setCreateIfMissing(true).setEncrypt(false).setKvStoreType(KvStoreType.SINGLE_VERSION);
6           singleKvStore = kvManager.getKvStore(CREATE, STORE_ID);
7       } catch (KvStoreException exception) {
8       }
9       return singleKvStore;
10  }
```

4. 订阅分布式数据库中数据变化

订阅分布式数据库中数据的变化需要实现 KvStoreObserver 接口、构造并注册 KvStoreObserver 实例。当被订阅的分布式数据库中的数据发生变化时,会调用 KvStoreObserverClient 中的 onChange()方法,在 onChange()方法中调用界面更新方法 updateAllData()把数据库中的数据显示到手机界面上,详细代码如程序清单 5-30 所示。

程序清单 5-30: hmos\ch05\02\DistributedDataServiceTest\entry\src\main\java\com\example\distributeddataservicetest\slice\MainAbilitySlice.java

```java
1   private void subscribeDb(SingleKvStore singleKvStore) {
2       KvStoreObserver kvStoreObserverClient = new KvStoreObserverClient();
3       singleKvStore.subscribe(SubscribeType.SUBSCRIBE_TYPE_ALL, kvStoreObserverClient);
4   }
5
6   private class KvStoreObserverClient implements KvStoreObserver {
7       @Override
8       public void onChange(ChangeNotification notification) {
9           getUITaskDispatcher().asyncDispatch(new Runnable() {
10              @Override
11              public void run() {
12                  updateAllData();
13              }
14          });
15      }
16  }
```

最后创建一个 initDbManager()方法来调用这些函数,并在 onStart()中调用 initDbManager()方法,详细代码如程序清单 5-31 所示。

程序清单 5-31: hmos\ch05\02\DistributedDataServiceTest\entry\src\main\java\com\example\distributeddataservicetest\slice\MainAbilitySlice.java

```
1    private void initDbManager() {
2        kvManager = createManager();
3        singleKvStore = createDb(kvManager);
4        subscribeDb(singleKvStore);
5    }
```

5. 对数据进行数据查询、插入和删除操作

(1) 数据查询,详细代码如程序清单 5-32 所示。

程序清单 5-32: hmos\ch05\02\DistributedDataServiceTest\entry\src\main\java\com\example\distributeddataservicetest\slice\MainAbilitySlice.java

```
1    private void queryAllData() {
2        List<Entry> entryList = singleKvStore.getEntries("");
3        hashMap.clear();
4        for (Entry entry : entryList) {
5            hashMap.put(entry.getKey(), entry.getValue().getString());
6        }
7    }
```

(2) 数据插入,详细代码如程序清单 5-33 所示。

程序清单 5-33: hmos\ch05\02\DistributedDataServiceTest\entry\src\main\java\com\example\distributeddataservicetest\slice\MainAbilitySlice.java

```
1    private void addData(String key, String value) {
2        if (key == null || key.isEmpty() || value == null || value.isEmpty()) {
3            return;
4        }
5        singleKvStore.putString(key, value);
6        showTip("写入成功!");
7    }
```

(3) 数据删除,详细代码如程序清单 5-34。

程序清单 5-34: hmos\ch05\02\DistributedDataServiceTest\entry\src\main\java\com\example\distributeddataservicetest\slice\MainAbilitySlice.java

```
1    private void deleteData(String key) {
```

```
2       if (key.isEmpty()) {
3           return;
4       }
5       singleKvStore.delete(key);
6       showTip("删除成功!");
7   }
```

6. 同步数据到其他设备

在进行数据同步之前,首先需要获取当前组网环境中的设备列表,然后指定同步方式(PULL_ONLY,PUSH_ONLY,PUSH_PULL)进行同步,以 PUSH_PULL 方式为例,详细代码如程序清单 5-35 所示。

程序清单 5-35: hmos\ch05\02\DistributedDataServiceTest\entry\src\main\java\com\example\distributeddataservicetest\slice\MainAbilitySlice.java

```
1   private void syncOthers() {
2       List<DeviceInfo> deviceInfoList = kvManager.getConnectedDevicesInfo(DeviceFilterStrategy.NO_FILTER);
3       List<String> deviceIdList = new ArrayList<>();
4       for (DeviceInfo deviceInfo : deviceInfoList) {
5           deviceIdList.add(deviceInfo.getId());
6       }
7       if (deviceIdList.size() == 0) {
8           showTip("组网失败!");
9           return;
10      }
11      singleKvStore.registerSyncCallback(new SyncCallback() {
12          @Override
13          public void syncCompleted(Map<String, Integer> map) {
14              getUITaskDispatcher().asyncDispatch(new Runnable() {
15                  @Override
16                  public void run() {
17                      updateAllData();
18                      showTip("同步成功!");
19                  }
20              });
21              singleKvStore.unRegisterSyncCallback();
22          }
23      });
24      singleKvStore.sync(deviceIdList, SyncMode.PUSH_ONLY);
25  }
```

7. 界面更新

实现将数据库里的所有数据显示在手机界面上,以便查看数据库的操作结果,详细代码如程序清单 5-36 所示。

程序清单 5-36：hmos\ch05\02\DistributedDataServiceTest\entry\src\main\java\com\example\distributeddataservicetest\slice\MainAbilitySlice.java

```
1   private void updateAllData() {
2       queryAllData();
3       String finalString = "";
4       for (HashMap.Entry<String, String> entry : hashMap.entrySet()) {
5           finalString = finalString + entry.getKey() + ":" + entry.getValue() + "; ";
6       }
7       text_output.setText(finalString);
8   }
```

8. 设置监听事件

创建一个 initEvent() 方法实现监听事件，并在 onStart() 方法中调用 initEvent() 方法，使得用户单击相应按钮后，能进行相应的数据添加、数据删除和数据同步，详细代码如程序清单 5-37 所示。

程序清单 5-37：hmos\ch05\02\DistributedDataServiceTest\entry\src\main\java\com\example\distributeddataservicetest\slice\MainAbilitySlice.java

```
1   private void initEvent() {
2       key_input = (TextField) findComponentById(ResourceTable.Id_key_input);
3       value_input = (TextField) findComponentById(ResourceTable.Id_value_input);
4       text_output = (Text) findComponentById(ResourceTable.Id_text_output);
5       input_button = (Button) findComponentById(ResourceTable.Id_input_button);
6       delete_button = (Button) findComponentById(ResourceTable.Id_delete_button);
7       sync_button = (Button) findComponentById(ResourceTable.Id_sync_button);
8       input_button.setClickedListener(new Component.ClickedListener() {
9           @Override
10          public void onClick(Component component) {
11              addData(key_input.getText(), value_input.getText());
12          }
13      });
14      delete_button.setClickedListener(new Component.ClickedListener() {
15          @Override
```

```
16          public void onClick(Component component) {
17              String deleteKey = key_input.getText();
18              deleteData(deleteKey);
19          }
20      });
21      sync_button.setClickedListener(new Component.ClickedListener() {
22          @Override
23          public void onClick(Component component) {
24              syncOthers();
25          }
26      });
27  }
```

9. 示例的完整代码

单版本分布式数据库开发案例的主要文件 MainAbilitySlice.java 的完整代码如程序清单 5-38 所示。

程序清单 5-38: hmos\ch05\02\DistributedDataServiceTest\entry\src\main\java\com\example\distributeddataservicetest\slice\MainAbilitySlice.java

```
1   public class MainAbilitySlice extends AbilitySlice {
2       private static final String STORE_ID = "test_db";
3       private KvManager kvManager;
4       private SingleKvStore singleKvStore;
5       private static final int SHOW_TIME = 1500;
6       private HashMap<String, String> hashMap = new HashMap<>();
7       private Button input_button, delete_button, sync_button;
8       private TextField key_input, value_input;
9       private Text text_output;
10
11      @Override
12      public void onStart(Intent intent) {
13          super.onStart(intent);
14          super.setUIContent(ResourceTable.Layout_ability_main);
15          initDbManager();
16          initEvent();
17      }
18
19      private void initEvent() {
20          key_input = (TextField) findComponentById(ResourceTable.Id_key_input);
21          value_input = (TextField) findComponentById(ResourceTable.Id_value_input);
```

```
22          text_output = (Text) findComponentById(ResourceTable.Id_text_
   output);
23          input_button = (Button) findComponentById(ResourceTable.Id_
   input_button);
24          delete_button = (Button) findComponentById(ResourceTable.Id_
   delete_button);
25          sync_button = (Button) findComponentById(ResourceTable.Id_sync_
   button);
26          input_button.setClickedListener(new Component.ClickedListener() {
27              @Override
28              public void onClick(Component component) {
29                  addData(key_input.getText(), value_input.getText());
30              }
31          });
32          delete_button.setClickedListener(new Component.ClickedListener() {
33              @Override
34              public void onClick(Component component) {
35                  String deleteKey = key_input.getText();
36                  deleteData(deleteKey);
37              }
38          });
39          sync_button.setClickedListener(new Component.ClickedListener() {
40              @Override
41              public void onClick(Component component) {
42                  syncOthers();
43              }
44          });
45      }
46
47      private void initDbManager() {
48          kvManager = createManager();
49          singleKvStore = createDb(kvManager);
50          subscribeDb(singleKvStore);
51      }
52
53      private KvManager createManager() {
54          KvManager kvManager = null;
55          try {
56              KvManagerConfig kvManagerConfig = new KvManagerConfig(this);
57              kvManager = KvManagerFactory.getInstance().createKvManager
   (kvManagerConfig);
58          } catch (KvStoreException exception) {
59          }
```

```
60              return kvManager;
61          }
62
63      private SingleKvStore createDb(KvManager kvManager) {
64          SingleKvStore singleKvStore = null;
65          try {
66              Options CREATE = new Options();
67              CREATE.setCreateIfMissing(true).setEncrypt(false).setKvStoreType(KvStoreType.SINGLE_VERSION);
68              singleKvStore = kvManager.getKvStore(CREATE, STORE_ID);
69          } catch (KvStoreException exception) {
70          }
71          return singleKvStore;
72      }
73
74      private void subscribeDb(SingleKvStore singleKvStore) {
75          KvStoreObserver kvStoreObserverClient = new KvStoreObserverClient();
76          singleKvStore.subscribe(SubscribeType.SUBSCRIBE_TYPE_ALL, kvStoreObserverClient);
77      }
78
79      private class KvStoreObserverClient implements KvStoreObserver {
80          @Override
81          public void onChange(ChangeNotification notification) {
82              getUITaskDispatcher().asyncDispatch(new Runnable() {
83                  @Override
84                  public void run() {
85                      updateAllData();
86                  }
87              });
88          }
89      }
90
91      private void updateAllData() {
92          queryAllData();
93          String finalString = "";
94          for (HashMap.Entry<String, String> entry : hashMap.entrySet()) {
95              finalString = finalString + entry.getKey() + ":" + entry.getValue() + "; ";
96          }
97          text_output.setText(finalString);
98      }
```

```
 99
100     private void queryAllData() {
101         List<Entry> entryList = singleKvStore.getEntries("");
102         hashMap.clear();
103         for (Entry entry : entryList) {
104             hashMap.put(entry.getKey(), entry.getValue().getString());
105         }
106     }
107
108     private void addData(String key, String value) {
109         if (key == null || key.isEmpty() || value == null || value.isEmpty()) {
110             return;
111         }
112         singleKvStore.putString(key, value);
113         showTip("写入成功!");
114     }
115
116     private void deleteData(String key) {
117         if (key.isEmpty()) {
118             return;
119         }
120         singleKvStore.delete(key);
121         showTip("删除成功!");
122     }
123
124     /**
125      * 删除数据库
126      */
127     private void deleteDB() {
128         kvManager.closeKvStore(singleKvStore);
129         kvManager.deleteKvStore(STORE_ID);
130     }
131
132     private void syncOthers() {
133         List<DeviceInfo> deviceInfoList = kvManager.
    getConnectedDevicesInfo(DeviceFilterStrategy.NO_FILTER);
134         List<String> deviceIdList = new ArrayList<>();
135         for (DeviceInfo deviceInfo : deviceInfoList) {
136             deviceIdList.add(deviceInfo.getId());
137         }
138         if (deviceIdList.size() == 0) {
```

```
139                showTip("组网失败!");
140                return;
141            }
142            singleKvStore.registerSyncCallback(new SyncCallback() {
143                @Override
144                public void syncCompleted(Map<String, Integer> map) {
145                    getUITaskDispatcher().asyncDispatch(new Runnable() {
146                        @Override
147                        public void run() {
148                            updateAllData();
149                            showTip("同步成功!");
150                        }
151                    });
152                    singleKvStore.unRegisterSyncCallback();
153                }
154            });
155            singleKvStore.sync(deviceIdList, SyncMode.PUSH_ONLY);
156        }
157
158        private void showTip(String message) {
159            new ToastDialog(this).setAlignment(LayoutAlignment.CENTER)
160                    .setText(message).setDuration(SHOW_TIME).show();
161        }
162
163        @Override
164        public void onActive() {
165            super.onActive();
166        }
167
168        @Override
169        public void onForeground(Intent intent) {
170            super.onForeground(intent);
171        }
172    }
```

10. 测试与运行效果

（1）在远程设备中选择 Super device 下的两台 P40 系列远程设备的组合，如图 5-4 所示。

（2）分别在两台 P40 远程设备上安装项目，如图 5-5 所示。

（3）分别在两台 P40 远程设备上打开项目，如图 5-6 所示。

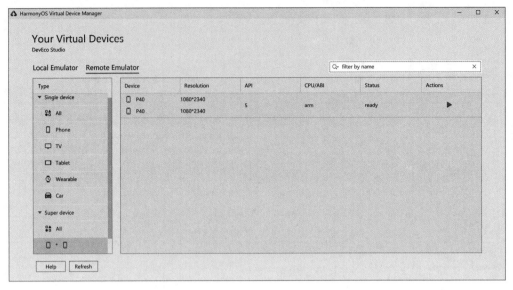

图 5-4　选择 Super device 下的两台 P40 系列远程设备的组合

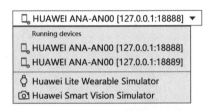

图 5-5　分别在两台 P40 远程设备上安装项目

图 5-6　分别在两台 P40 远程设备上打开项目

（4）输入任意值并单击"写入"按钮，相应的数据会被写入数据库中并且同步到另一台设备中，如图 5-7 所示。

图 5-7　写入数据显示效果

（5）单击"删除"按钮可删除相应的数据，此时删除的是刚刚写入的 key 值为 yd 的数据，删除成功后另一台设备上也会相应地删除该数据并更新界面，如图 5-8 所示。

图 5-8　删除数据显示效果

5.3 本章小结

本章主要讲解了 HarmonyOS 中基本的数据存储和传输的方法。HarmonyOS 内置了一个小型的基于 SQLite 的关系数据库,且为访问该关系数据库提供了大量方便的工具类。为了方便不同设备之间的数据共享,HarmonyOS 也为开发者提供了分布式数据服务,在通过可认证的设备间,为用户提供在多种终端设备上最终一致的数据访问体验。

5.4 课后习题

1. 在开发 HarmonyOS 应用程序时,如果希望在本地存储一些结构化的数据,可以使用(　　)。
   A. HarmonyOS 关系数据库　　　　　　B. MySQL
   C. DB2　　　　　　　　　　　　　　D. Sybase
2. HarmonyOS 关系数据库是基于 SQLite 的,以下数据类型中,(　　)不是 HarmonyOS 关系数据库内部支持的类型。
   A. NULL　　　　B. INTEGER　　　　C. STRING　　　　D. TEXT
3. 构造要插入 HarmonyOS 关系数据库的数据,是以(　　)形式存储的。
   A. xml　　　　　　　　　　　　　　B. txt
   C. ValuesBucket　　　　　　　　　　D. 根据用户自定义
4. 分布式数据服务的数据模型仅支持(　　)。
   A. 面向对象数据模型　　　　　　　　B. KV 数据模型
   C. XML 数据模型　　　　　　　　　　D. 函数数据模型
5. 在 4.2.8 节例子的基础上实现注册功能后,将信息保存到本地关系数据库;跳转到注册成功界面后,从关系数据库读取信息然后显示出来。

第 6 章

公共事件、通知与日志

课程思政 6

本章要点

- 四种公共事件的创建和使用
- 通知的创建和使用
- 日志的使用

本章知识结构图(见图 6-1)

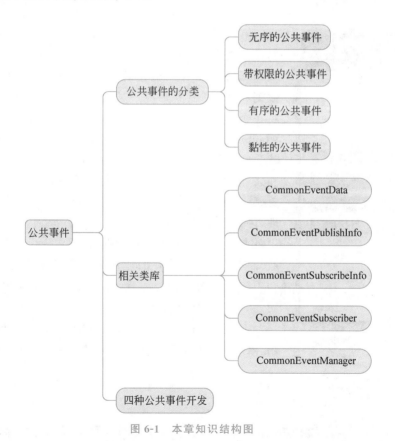

图 6-1 本章知识结构图

图 6-1（续）

本章示例（见图 6-2）

图 6-2　本章示例图

第 4 章中介绍了 Service Ability，可以将一些比较耗时的操作放在 Service Ability 中执行，通过调用相应的方法获取 Service Ability 中数据的状态；如果需要得到某一特定的数据状态，则需要每隔一段时间调用一次该方法，然后判断是否达到想要的状态，这样非常不方便。如果 Service Ability 能够在数据状态满足一定条件时主动给出通知，那就非常人性化了。HarmonyOS 中提供了公共事件的订阅和发布功能，本质上它就是一个全局的监听器，一直监听着某一消息，一旦收到该消息，则触发相应的方法进行处理，因此可以非常方便地在不同组件间通信。最典型的应用就是电量提醒，当手机电量低于某一设定值时，则发出通知信息。一个较好的应用程序也要能够通过通知栏通知用户应用内的相关消息，例如 QQ 和微信的新消息提示功能。

本章将通过一个综合案例讲解公共事件以及通知的使用，最后将简单介绍之前章节中偶尔使用到的日志功能。

6.1　公共事件

HarmonyOS 通过 CES（Common Event Service，公共事件服务）为应用程序提供订阅、发布、退订公共事件的能力。每个应用都可以订阅自己感兴趣的公共事件，订阅成功且公共事件发布后，系统会把其发送给应用。这些公共事件可能来自系统、其他应用和应用自身。

公共事件可分为系统公共事件和自定义公共事件。系统公共事件是系统将收集到的事件信息,根据系统策略发送给订阅该事件的用户程序,例如亮灭屏事件、USB 插拔、网络连接、系统升级等。自定义公共事件是开发者自定义一些用来处理业务逻辑的公共事件。应用如果需要接收公共事件,则需要订阅相应的事件。

6.1.1 四种公共事件

开发者可以发布四种公共事件:无序的公共事件、带权限的公共事件、有序的公共事件、黏性的公共事件。

- 无序的公共事件:最基础的公共事件,没有任何限制条件。
- 带权限的公共事件:当订阅的事件发生了,会对订阅者做一个筛选,并不是让所有的订阅者都做出响应。
- 有序的公共事件:多个订阅者有依赖关系或者对处理顺序有要求,例如高优先级订阅者可修改公共事件内容或处理结果,包括终止公共事件处理;或者低优先级订阅者依赖高优先级的处理结果等。
- 黏性的公共事件:当公共事件的订阅动作在公共事件发布之后发生时,订阅者也能接收到公共事件并做出响应。

6.1.2 公共事件相关类

公共事件相关类包含 CommonEventData、CommonEventPublishInfo、CommonEventSubscribeInfo、CommonEventSubscriber 和 CommonEventManager,它们之间的关系如图 6-3 所示。

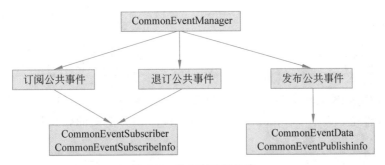

图 6-3 公共事件相关类

CommonEventData 封装公共事件相关信息,用于在发布、分发和接收时处理数据。CommonEventData 类定义的常用操作方法如表 6-1 所示。

表 6-1 CommonEventData 类定义的常用操作方法

方法	描述
CommonEventData()	创建公共事件数据
CommonEventData(Intent intent)	创建公共事件数据指定 Intent
CommonEventData(Intent intent, int code, String data)	创建公共事件数据,指定 Intent、code 和 data

续表

方 法	描 述
Intent getIntent()	获取公共事件 Intent
void setCode(int code)	设置有序公共事件的结果码
int getCode()	获取有序公共事件的结果码
int setData(String data)	设置有序公共事件的详细结果数据
String getData()	获取有序公共事件的详细结果数据

CommonEventPublishInfo 封装公共事件发布相关属性、限制等信息,包括公共事件的类型、接收者的权限等。CommonEventPublishInfo 类定义的常用操作方法如表 6-2 所示。

表 6-2 CommonEventPublishInfo 类定义的常用操作方法

方 法	描 述
CommonEventPublishInfo()	创建公共事件信息
CommonEventPublishInfo(CommonEventPublishInfo publishInfo)	复制一个公共事件信息
void setSticky(boolean sticky)	设置公共事件的黏性属性
boolean isSticky()	获取它是否为黏性公共事件
void setOrdered(boolean ordered)	设置公共事件的有序属性
boolean isOrdered()	获取它是否为有序公共事件
void setSubscriberPermissions(String[] subscriberPermissions)	设置公共事件订阅者的权限,若为多个参数,则仅第一个参数生效

CommonEventSubscribeInfo 封装公共事件订阅相关信息,比如公共事件的优先级、线程模式、事件范围等。CommonEventSubscribeInfo 类定义的常用操作方法如表 6-3 所示。

表 6-3 CommonEventSubscribeInfo 类定义的常用操作方法

方 法	描 述
CommonEventSubscribeInfo(MatchingSkills matchingSkills)	创建公共事件订阅器指定 matchingSkills
MatchingSkills getMatchingSkills()	获取此 CommonEventSubscriptInfo 对象中携带的 MatchingSkills 对象
CommonEventSubscribeInfo(CommonEventSubscribeInfo)	复制公共事件订阅器对象
void setPriority(int priority)	设置优先级,用于有序公共事件
int getPriority()	获取此 CommonEventSubscribeInfo 对象的优先级

续表

方法	描述
void setThreadMode(ThreadMode threadMode)	指定订阅者的回调函数运行在哪个线程上
ThreadMode getThreadMode()	获取此 CommonEventSubscribeInfo 对象的线程
void setPermission(String permission)	设置发布者必须具备的权限
String getPermission()	获取此 CommonEventSubscribeInfo 对象的发布者具备的权限
void setDeviceId(String deviceId)	指定订阅哪台设备的公共事件
String getDeviceId()	获取此 CommonEventSubscribeInfo 对象中指定的设备 ID

CommonEventSubscriber 封装公共事件订阅者及相关参数。CommonEventSubscriber 类定义的常用操作方法如表 6-4 所示。

表 6-4　CommonEventSubscriber 类定义的常用操作方法

方法	描述
CommonEventSubscriber(CommonEventSubscribeInfosubscribeInfo)	构造公共事件订阅者实例
abstract void onReceiveEvent(CommonEventData data)	由开发者实现，在接收到公共事件时被调用
AsyncCommonEventResultgoAsyncCommonEvent()	设置有序公共事件异步执行
void setCodeAndData(int code, String data)	设置有序公共事件的异步结果
void setData(String data)	设置有序公共事件的异步结果数据
String getData()	获取有序公共事件的异步结果数据
void setCode(int code)	设置有序公共事件的异步结果码
int getCode()	获取有序公共事件的异步结果码
void setCodeAndData(int code, String data)	设置有序公共事件的异步结果码和异步结果数据
void abortCommonEvent()	取消当前的公共事件，仅对有序公共事件有效，取消后，公共事件不再向下一个订阅者传递
boolean getAbortCommonEvent()	获取当前有序公共事件是否取消的状态
void clearAbortCommonEvent()	清除当前有序公共事件的 abort 状态
boolean isOrderedCommonEvent()	查询当前公共事件是否为有序公共事件
boolean isStickyCommonEvent()	查询当前公共事件是否为黏性公共事件

CommonEventManager 是为应用提供订阅、退订和发布公共事件的静态接口类。CommonEventManager 类定义的常用操作方法如表 6-5 所示。

表 6-5　CommonEventManager 类定义的常用操作方法

方　　法	描　　述
static void publishCommonEvent(CommonEventData eventData)	发布公共事件
static void publishCommonEvent（CommonEventData event，CommonEventPublishInfo publishInfo)	发布公共事件指定发布信息
static void publishCommonEvent（CommonEventData event，CommonEventPublishInfo publishInfo，CommonEventSubscriber resultSubscriber)	发布有序公共事件，指定发布信息和最后一个接收者
static void subscribeCommonEvent（CommonEventSubscriber subscriber)	订阅公共事件
static void unsubscribeCommonEvent（CommonEventSubscriber subscriber)	退订公共事件

6.1.3　无序的公共事件开发

首先是公共事件的发布。发布公共事件通过构造 CommonEventData 对象，设置 Intent，再通过构造 operation 对象把需要发布的公共事件信息传入 intent 对象；然后调用 CommonEventManager.publishCommonEvent(CommonEventData) 接口发布公共事件，详细代码如程序清单 6-1 所示。

程序清单 6-1：hmos\ch06\01\CommonEventServiceTest\entry\src\main\java\com\example\commoneventservicetest\slice\MainAbilitySlice.java

```
1   private void Mypublish(){
2       System.out.println("发布事件开始了");
3       Intent intent = new Intent();
4       Operation operation = new Intent.OperationBuilder()
5               .withAction("action.event")              //定义的事件的标识符
6               .build();
7       intent.setOperation(operation);
8       //构建 CommonEventData 对象
9       CommonEventData commonEventData = new CommonEventData(intent);
10      try {
11          CommonEventManager.publishCommonEvent(commonEventData);//无序的
12          System.out.println("发布事件完成了");
13      } catch (RemoteException e) {
14          e.printStackTrace();
15      }
16  }
```

接下来是订阅公共事件。先新建一个 subscriber 包专门用来放有关订阅者的类，在该包下创建一个 CommonEventSubscriber 派生类，在 onReceiveEvent()回调函数中处理公共事件。这里需要注意,在 onReceiveEvent()函数中不能执行耗时操作,可以使用 CommonEventSubscriber 的 goAsyncCommonEvent()方法实现异步操作,函数返回后仍保持该公共事件活跃,且执行完成后必须调用 finishCommonEvent()方法结束,详细代码如程序清单 6-2 所示。

程序清单 6-2：hmos\ch06\01\CommonEventServiceTest\entry\src\main\java\com\example\commoneventservicetest\subscriber\MyCommonEventSubscriber1.java

```java
1   public class MyCommonEventSubscriber1 extends CommonEventSubscriber {
2       public MyCommonEventSubscriber1(CommonEventSubscribeInfo subscribeInfo) {
3           super(subscribeInfo);
4       }
5
6       EventHandler eventHandler = new EventHandler(EventRunner.create());
7
8       @Override
9       public void onReceiveEvent(CommonEventData commonEventData) {
10          System.out.println("不耗时的任务开始 1");
11          //耗时任务应该在子线程中执行
12          AsyncCommonEventResult asyncCommonEventResult = goAsyncCommonEvent();
13          eventHandler.postTask(new Runnable() {
14              @Override
15              public void run() {
16                  try {
17                      Thread.sleep(2000);
18                  } catch (InterruptedException e) {
19                      e.printStackTrace();
20                  }
21                  asyncCommonEventResult.finishCommonEvent();
                    //上一个订阅者的任务没完成之前,阻止事件传递到下一个订阅者
22                  System.out.println("耗时的响应事件的任务开始 1");
23              }
24          });
25      }
26  }
```

接下来构造 MyCommonEventSubscriber1 对象,通过调用 CommonEventManager.subscribeCommonEvent()方法进行订阅。需要注意的是,同一公共事件不能重复订阅。这里我们订阅 Mypublish()方法中自定义的公共事件,详细代码如程序清单 6-3 所示。

程序清单 6-3：hmos\ch06\01\CommonEventServiceTest\entry\src\main\java\com\example\commoneventservicetest\slice\MainAbilitySlice.java

```java
1   private void Mysubscribe() {
2       System.out.println("订阅事件开始");
3       //构建 MatchingSkills 对象
4       MatchingSkills matchingSkills = new MatchingSkills();
5       matchingSkills.addEvent("action.event");
                                        //订阅事件,可以在这里订阅系统公共事件
6       //构建订阅信息对象
7       CommonEventSubscribeInfo commonEventSubscribeInfo = new CommonEventSubscribeInfo(matchingSkills);
8       //构建订阅者对象
9       subscriber1 = new MyCommonEventSubscriber1(commonEventSubscribeInfo);
10      try {
11          CommonEventManager.subscribeCommonEvent(subscriber1);
12      } catch (RemoteException e) {
13          e.printStackTrace();
14      }
15  }
```

对于退订公共事件，可调用 CommonEventManager.unsubscribeCommonEvent()方法实现。通常在 onStop()中编写退订公共事件的代码，这样可以达到 Ability 关掉后自动取消订阅事件的目的。调用后，之前订阅的所有公共事件均被退订，详细代码如程序清单 6-4 所示。

程序清单 6-4：hmos\ch06\01\CommonEventServiceTest\entry\src\main\java\com\example\commoneventservicetest\slice\MainAbilitySlice.java

```java
1   private void Myunsubscribe() {
2       try {
3           CommonEventManager.unsubscribeCommonEvent(subscriber1);
4           System.out.println("事件退订成功");
5       } catch (RemoteException e) {
6           e.printStackTrace();
7       }
8   }
```

编写好相应的方法后，在 MainAbilitySlice 的 onStart()方法中调用这些方法。注意，对于无序的公共事件，需要先订阅，然后才能在相应的事件发布后响应，所以要先调用 Mysubscribe()方法，订阅事件后再调用 Mypublish()方法发布事件，代码演示效果如图 6-4 所示。

图 6-4　无序的公共事件案例日志输出

6.1.4　带权限的公共事件开发

在之前编写的无序的公共事件开发代码的基础上学习带权限的公共事件开发。在编写发布公共事件代码之前，先在配置文件（config.json）中的"module＞abilities"字段下声明一个自定义的权限。详细代码如程序清单 6-5 所示。

程序清单 6-5：hmos\ch06\01\CommonEventServiceTest\entry\src\main\config.json

```
1   "defPermissions": [
2     {
3       "name": "com.example.commoneventservicetest.permission",
4       "grantMode": "system_grant",
5       "availableScope": "signature"
6     }
7   ]
```

声明完权限后，再来编写发布公共事件的 Mypublish()方法，发布有权限的公共事件的代码大体上与无序的公共事件中的一致，利用 CommonEventPublishInfo 类和其相关接口绑定之前在配置文件中声明的权限，详细代码如程序清单 6-6 所示。

程序清单 6-6：hmos\ch06\01\CommonEventServiceTest\entry\src\main\
java\com\example\commoneventservicetest\slice\MainAbilitySlice.java

```
1   private void Mypublish() {
2       System.out.println("发布事件开始了");
3       Intent intent = new Intent();
4       Operation operation = new Intent.OperationBuilder()
5               .withAction("action.event")    //定义事件的标识符
6               .build();
7       intent.setOperation(operation);
8       //构建 CommonEventData 对象
9       CommonEventData commonEventData = new CommonEventData(intent);
10      CommonEventPublishInfo commonEventPublishInfo = new CommonEventPublishInfo();
```

```
11      String[] permissions = {"com.example.commoneventservicetest.
    permission"};
12      commonEventPublishInfo.setSubscriberPermissions(permissions);
13      try {
14          CommonEventManager.publishCommonEvent(commonEventData,
    commonEventPublishInfo);                      //设置权限
15          System.out.println("发布事件完成了");
16      } catch (RemoteException e) {
17          e.printStackTrace();
18      }
19  }
```

给订阅者所关联的 MainAbility 设置响应公共事件的权限,在配置文件(config.json)中的"module>abilities"字段下声明权限。声明好权限后,就编写用于订阅公共事件的函数,该函数与无序的公共事件的函数一致。声明权限的详细代码如程序清单6-7所示。

程序清单 6-7：hmos\ch06\01\CommonEventServiceTest\entry\src\main\config.json

```
1   "reqPermissions": [
2   {
3     "name": "com.example.commoneventservicetest.permission",
4     "reason": "get right",
5     "usedScene": {
6       "ability": [
7         ".MainAbility"
8       ],
9       "when": "inuse"
10    }
11  }
12  ]
```

同样,在 MainAbilitySlice 的 onStart()方法中按先订阅后发布的顺序调用这些方法,代码演示效果如图6-5所示。

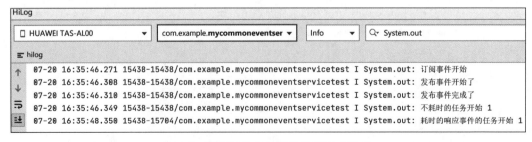

图 6-5　带权限的公共事件案例日志输出

6.1.5 有序的公共事件开发

在无序的公共事件开发的基础再新建一个 CommonEventSubscriber 派生类,把它命名为 MyCommonEventSubscriber2,里面的函数和代码复制 MyCommonEventSubscriber1 中的,详细代码如程序清单 6-8 所示。

程序清单 6-8:hmos\ch06\01\CommonEventServiceTest\entry\src\main\java\com\example\commoneventservicetest\subscriber\MyCommonEventSubscriber2.java

```
1   public class MyCommonEventSubscriber2 extends CommonEventSubscriber {
2       public MyCommonEventSubscriber2(CommonEventSubscribeInfo subscribeInfo) {
3           super(subscribeInfo);
4       }
5
6       EventHandler eventHandler = new EventHandler(EventRunner.create());
7
8       @Override
9       public void onReceiveEvent(CommonEventData commonEventData) {
10          System.out.println("不耗时的任务开始 2");
11          //耗时任务应该在子线程中执行
12          AsyncCommonEventResult asyncCommonEventResult = goAsyncCommonEvent();
13          eventHandler.postTask(new Runnable() {
14              @Override
15              public void run() {
16                  try {
17                      Thread.sleep(2000);
18                  } catch (InterruptedException e) {
19                      e.printStackTrace();
20                  }
21                  asyncCommonEventResult.finishCommonEvent();
                    //上一个订阅者的任务没完成之前,阻止事件传递到下一个订阅者
22                  System.out.println("耗时的响应事件的任务开始 2");
23              }
24          });
25      }
26  }
```

然后编写发布有序的公共事件的 Mypublish() 方法,该处的 Mypublish() 方法大体上与无序的公共事件中的一致,我们通过 CommonEventPublishInfo.setOrdered(true) 设置该公共事件是有序的公共事件,详细代码如程序清单 6-9 所示。

程序清单 6-9：hmos\ch06\01\CommonEventServiceTest\entry\src\main\java\com\example\commoneventservicetest\slice\MainAbilitySlice.java

```java
private void Mypublish() {
    System.out.println("发布事件开始了");
    Intent intent = new Intent();
    Operation operation = new Intent.OperationBuilder()
            .withAction("action.event")      //定义的事件的标识符
            .build();
    intent.setOperation(operation);
    //构建 CommonEventData 对象
    CommonEventData commonEventData = new CommonEventData(intent);
    CommonEventPublishInfo commonEventPublishInfo = new CommonEventPublishInfo();
    commonEventPublishInfo.setOrdered(true);
    try {
        CommonEventManager.publishCommonEvent(commonEventData, commonEventPublishInfo);                //设置权限
        System.out.println("发布事件完成了");
    } catch (RemoteException e) {
        e.printStackTrace();
    }
}
```

之后再编写订阅有序的公共事件的 Mysubscribe() 方法，该处的 Mysubscribe() 方法大体上与无序的公共事件中的一致，我们需要实例化两个订阅者信息，并通过 CommonEventSubscribeInfo.setPriority() 设置两个订阅者的优先级，优先级的取值范围为-1000～1000，默认为 0，值越大优先级越高，详细代码如程序清单 6-10 所示。

程序清单 6-10：hmos\ch06\01\CommonEventServiceTest\entry\src\main\java\com\example\commoneventservicetest\slice\MainAbilitySlice.java

```java
private void Mysubscribe() {
    System.out.println("订阅事件开始");
    //构建 MatchingSkills 对象
    MatchingSkills matchingSkills = new MatchingSkills();
    matchingSkills.addEvent("action.event");
    //构建订阅信息对象
    CommonEventSubscribeInfo commonEventSubscribeInfo1=new CommonEventSubscribeInfo(matchingSkills);
    CommonEventSubscribeInfo commonEventSubscribeInfo2=new CommonEventSubscribeInfo(matchingSkills);
```

```
 9      commonEventSubscribeInfo1.setPriority(0);      //设置优先级
10      commonEventSubscribeInfo2.setPriority(100);
11      //构建订阅者对象
12      subscriber1 = new MyCommonEventSubscriber1
        (commonEventSubscribeInfo1);
13      subscriber2 = new MyCommonEventSubscriber2
        (commonEventSubscribeInfo2);
14      try {
15          CommonEventManager.subscribeCommonEvent(subscriber1);
16          CommonEventManager.subscribeCommonEvent(subscriber2);
17      } catch (RemoteException e) {
18          e.printStackTrace();
19      }
20  }
```

同样，在 MainAbilitySlice 的 onStart()方法中按先订阅后发布的顺序调用这些方法，代码演示效果如图 6-6 所示。

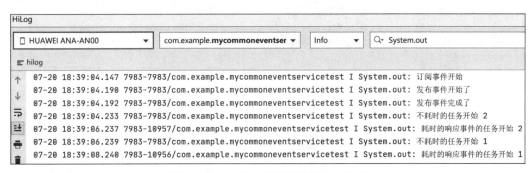

图 6-6 有序的公共事件案例日志输出

6.1.6 黏性的公共事件开发

黏性的公共事件涉及系统的权限，需要在配置文件（config.json）中的"module＞abilities"字段下声明 ohos.permission.COMMONEVENT_STICKY 权限，声明权限的详细代码如程序清单 6-11 所示。

程序清单 6-11：hmos\ch06\01\CommonEventServiceTest\entry\src\main\config.json

```
1  "reqPermissions": [
2    {
3      "name": "ohos.permission.COMMONEVENT_STICKY",
4      "reason": "Obtain the required permission",
5      "usedScene": {
6        "ability": [
```

```
7            ".MainAbility"
8        ],
9        "when": "inuse"
10    }
11  }
12 ]
```

然后编写发布黏性的公共事件的 Mypublish() 方法，该处的 Mypublish() 方法与无序的公共事件中的方法大体上一致，我们通过 CommonEventPublishInfo.setSticky(true) 设置该公共事件是黏性的公共事件。而订阅黏性的公共事件的 Mysubscribe() 方法与订阅有序的公共事件中的完全一致，详细代码如程序清单 6-12 所示。

程序清单 6-12：hmos\ch06\01\CommonEventServiceTest\entry\src\main\java\com\example\commoneventservicetest\slice\MainAbilitySlice.java

```
1  private void Mypublish() {
2      System.out.println("发布事件开始了");
3      Intent intent = new Intent();
4      Operation operation = new Intent.OperationBuilder()
5              .withAction("action.event")        //定义的事件的标识符
6              .build();
7      intent.setOperation(operation);
8      //构建 CommonEventData 对象
9      CommonEventData commonEventData = new CommonEventData(intent);
10     CommonEventPublishInfo commonEventPublishInfo = new CommonEventPublishInfo();
11     commonEventPublishInfo.setSticky(true);    //设置成黏性的
12     //核心的发布事件的动作
13     try {
14         CommonEventManager.publishCommonEvent(commonEventData,
   commonEventPublishInfo);                      //设置权限
15         System.out.println("发布事件完成了");
16     } catch (RemoteException e) {
17         e.printStackTrace();
18     }
19 }
```

这一次，在 MainAbilitySlice 的 onStart() 方法中按先发布后订阅的顺序调用这些方法，再来看控制台的打印结果。与之前不同的是，这次先执行的是事件发布操作，当再次执行完订阅事件操作后，应用同样会响应事件的发生，结果如图 6-7 所示。

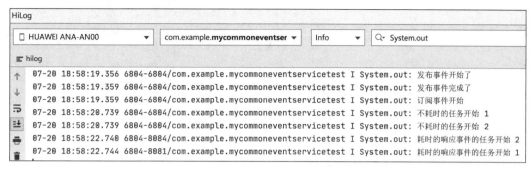

图 6-7　黏性的公共事件案例日志输出

6.2　通知

HarmonyOS 还给开发者提供了通知功能，即在手机的下滑通知栏内显示自定义的消息，主要用来提醒用户有来自该应用中的信息，例如我们常用的 QQ 及微信都会在手机的下滑通知中显示最新的消息。当应用向系统发出通知时，它将先以图标的形式显示在通知栏中，用户可以下拉通知栏查看通知的详细信息。

6.2.1　通知相关类

通知相关基础类包含 NotificationSlot、NotificationRequest 和 NotificationHelper，它们之间的关系如图 6-8 所示。

图 6-8　通知相关类

NotificationSlot 可以设置公共通知的振动、锁屏模式、重要级别等。一个应用可以创建一个或多个 NotificationSlot，在发布通知时，通过绑定不同的 NotificationSlot 实现不同的用途。NotificationSlot 类定义的常用操作方法如表 6-6 所示。

表 6-6　NotificationSlot 类定义的常用操作方法

方　　法	描　　述
NotificationSlot(String id，String name，int level)	构造 NotificationSlot
void setLevel(int level)	设置 NotificationSlot 的级别

续表

方　法	描　述
void setName(String name)	设置 NotificationSlot 的命名
void setDescription(String description)	设置 NotificationSlot 的描述信息
void enableBypassDnd(boolean bypassDnd)	设置是否绕过系统的免打扰模式
void setEnableVibration(boolean vibration)	设置收到通知时是否使能振动
void setEnableLight(boolean isLightEnabled)	设置收到通知时是否开启呼吸灯,前提是当前硬件支持呼吸灯
setLedLightColor(int color)	设置收到通知时的呼吸灯颜色
setLockscreenVisibleness(int visibleness)	设置在锁屏场景下,收到通知后是否显示,以及显示的效果

NotificationSlot 的级别目前支持 5 种,由低到高分别为
- LEVEL_NONE：表示通知不发布。
- LEVEL_MIN：表示通知可以发布,但是不显示在通知栏,不自动弹出,无提示音；该级别不适用于前台服务的场景。
- LEVEL_LOW：表示通知可以发布且显示在通知栏,不自动弹出,无提示音。
- LEVEL_DEFAULT：表示通知发布后可在通知栏显示,不自动弹出,触发提示音。
- LEVEL_HIGH：表示通知发布后可在通知栏显示,自动弹出,触发提示音。

NotificationRequest 用于设置具体的通知对象,包括设置通知的属性,如通知的分发时间、小图标、大图标、自动删除等参数,以及设置具体的通知类型,如普通文本、长文本等。NotificationRequest 类定义的常用操作方法如表 6-7 所示。

表 6-7　NotificationRequest 类定义的常用操作方法

方　法	描　述
NotificationRequest()	构建一个通知
NotificationRequest(int notificationId)	构建一个通知,指定通知的 ID。通知的 ID 在应用内容中具有唯一性,如果不指定,默认为 0
setNotificationId(int notificationId)	设置当前通知为 ID
setAutoDeletedTime(long time)	设置通知自动取消的时间戳
setContent(NotificationContent content)	设置通知的具体内容
setDeliveryTime(long deliveryTime)	设置通知分发的时间戳
setSlotId(String slotId)	设置通知的 NotificationSlot id
setTapDismissed(boolean tapDismissed)	设置通知在用户单击后是否自动取消
setShowDeliveryTime(boolean deliveryTime)	设置是否显示此通知的通知传递时间
setIntentAgent(IntentAgent agent)	设置通知承载指定的 IntentAgent,在通知中实现即将触发的事件

NotificationHelper 封装了发布、更新、删除通知等静态方法。NotificationHelper 类提供的方法都是静态的，可以在不创建 NotificationHelper 类对象的情况下调用。NotificationHelper 类定义的常用操作方法如表 6-8 所示。

表 6-8 NotificationHelper 类定义的常用操作方法

方 法	描 述
publishNotification(NotificationRequest request)	发布一条通知
cancelNotification(int notificationId)	取消指定的通知
cancelAllNotifications()	取消之前发布的所有通知
addNotificationSlot(NotificationSlot slot)	创建一个 NotificationSlot

6.2.2 通知开发示例

通知的开发具体分为以下 4 个步骤。
(1) 创建 NotificationSlot，设置通知的振动、锁屏模式、重要级别等。
(2) 构建 NotificationRequest 对象，绑定 NotificationSlot 对象。
(3) 调用 NotificationRequest 对象的 setContent() 设置通知的内容和类型。
(4) 调用 NotificationHelper 的 publishNotification() 发布通知。
发布通知的详细代码如程序清单 6-13 所示。

程序清单 6-13：hmos\ch06\01\CommonEventServiceTest\entry\src\main\java\com\example\commoneventservicetest\slice\MainAbilitySlice.java

```
1   private void publishNotification() {
2       //1.创建 NotificationSlot,设置通知的振动、锁屏模式、重要级别等
3       NotificationSlot slot = new NotificationSlot("slot_001", "slot_default", NotificationSlot.LEVEL_MIN);       //创建 NotificationSlot 对象
4       slot.setDescription("NotificationSlotDescription");      //设置描述信息
5       slot.setEnableVibration(true);                            //设置振动提醒
6       slot.setLockscreenVisibleness(NotificationRequest.VISIBLENESS_TYPE_PUBLIC);                                //设置锁屏模式
7       slot.setEnableLight(true);                                //设置开启呼吸灯提醒
8       slot.setLedLightColor(Color.RED.getValue());    //设置呼吸灯的提醒颜色
9       try {
10          NotificationHelper.addNotificationSlot(slot);
                          //发布 NotificationSlot 对象,后面使用了才能起作用
11      } catch (RemoteException e) {
12          e.printStackTrace();
13      }
14      //2.构建 NotificationRequest 对象,绑定 NotificationSlot 对象
15      int notificationId = 1;
```

```
16      NotificationRequest request = new NotificationRequest
    (notificationId);
17      request.setSlotId(slot.getId());
18      //3.调用 NotificationRequest 对象的 setContent()设置通知的内容和类型
19      NotificationRequest.NotificationNormalContent normalContent=new
    NotificationRequest.NotificationNormalContent();
20      normalContent.setTitle("通知标题").setText("通知内容").
    setAdditionalText("通知的次级内容");
21      NotificationRequest.NotificationContent notificationContent=new
    NotificationRequest.NotificationContent(normalContent);
22      request.setContent(notificationContent);              //设置通知的内容
23      request.setShowDeliveryTime(true);                    //通知后面显示时间
24      //4.调用 NotificationHelper 的 publishNotification()发布通知
25      try {
26          NotificationHelper.publishNotification(request);
27      } catch (RemoteException e) {
28          e.printStackTrace();
29      }
30  }
```

在 MainAbilitySlice 的 onStart()中调用上面创建好的 publishNotification()方法来发布通知,当应用运行成功时,会自动发布我们定义好的通知,如图 6-9 所示。

图 6-9 通知发布效果图

当然,通知可以自动发布,也可以自动取消。取消通知的方法有两种:一种是调用

NotificationHelper.cancelNotification(int notificationId)取消指定 ID 的通知；另一种是调用 NotificationHelper.cancelAllNotifications()取消所有本 App 发布的通知，详细代码如程序清单 6-14 所示。

程序清单 6-14：hmos\ch06\01\CommonEventServiceTest\entry\src\main\java\com\example\commoneventservicetest\slice\MainAbilitySlice.java

```
1    private void publishCancelNotification() {
2        try {
3            NotificationHelper.cancelNotification(1);    //取消指定通知
4            NotificationHelper.cancelAllNotifications();
                                                //取消所有本 App 发布的通知
5        } catch (RemoteException e) {
6            e.printStackTrace();
7        }
8    }
```

6.2.3 单击通知栏事件

当发布通知后，都希望使用者能够单击发布的通知，然后跳转到 App 中相应的界面。那么这该如何实现呢？HarmonyOS 提供了 IntentAgent 类来实现这一操作。IntentAgent 封装了一个指定行为的 Intent，可以通过 triggerIntentAgent 接口主动触发，也可以与通知绑定被动触发。IntentAgent 相关基础类包括 IntentAgentHelper、IntentAgentInfo 和 IntentAgentConstant。

IntentAgentHelper 封装了获取、取消 IntentAgent 等静态方法。IntentAgentHelper 类定义的常用操作方法如表 6-9 所示。

表 6-9 IntentAgentHelper 类定义的常用操作方法

方法	描述
staticIntentAgent getIntentAgent(Context context, IntentAgentInfo paramsInfo)	获取一个 IntentAgent 实例
static void cancel(IntentAgent agent)	取消一个 IntentAgent 实例
static boolean judgeEquality(IntentAgent agent, IntentAgent otherAgent)	判断两个 IntentAgent 实例是否相等

IntentAgentInfo 类封装了获取一个 IntentAgent 实例所需的数据。通常使用构造函数 IntentAgentInfo(int requestCode, OperationType operationType, List＜Flags＞ flags, List＜Intent＞ intents, IntentParams extraInfo)获取 IntentAgentInfo 对象。

- requestCode：请求码，使用者自定义。
- operationType：为 IntentAgentConstant.OperationType 枚举中的值。
- flags：为 IntentAgentConstant.Flags 枚举中的值。
- intents：将被执行的意图列表。

- extraInfo：表明如何启动一个有页面的 Ability，可以为 null，只在 operationType 的值为 START_ABILITY 和 START_ABILITIES 时有意义。

IntentAgentConstant 类中包含 OperationType 和 Flags 两个枚举类，如表 6-10 所示。

表 6-10 IntentAgentConstant 包含的枚举类

类 名	枚 举 值
IntentAgentConstant.OperationType	UNKNOWN_TYPE：不识别的类型
	START_ABILITY：开启一个有页面的 Ability
	START_ABILITIES：开启多个有页面的 Ability
	START_SERVICE：开启一个无页面的 Ability
	SEND_COMMON_EVENT：发送一个公共事件
IntentAgentConstant.Flags	ONE_TIME_FLAG：IntentAgent 仅能使用一次
	NO_BUILD_FLAG：如果描述 IntentAgent 对象不存在，则不创建它，直接返回 null
	CANCEL_PRESENT_FLAG：在生成一个新的 IntentAgent 对象前取消已存在的一个 IntentAgent 对象
IntentAgentConstant.Flags	UPDATE_PRESENT_FLAG：使用新的 IntentAgent 中的额外数据替换已存在的 IntentAgent 中的额外数据
	CONSTANT_FLAG：IntentAgent 是不可变的

下面介绍一个简单的示例。在 publishNotification()中的第三步添加代码，通过自定义的 getIntentAgent()方法获取 IntentAgent 对象，并调用 setIntentAgent()方法设置 IntentAgent，具体代码如程序清单 6-15 所示。

程序清单 6-15：hmos\ch06\01\CommonEventServiceTest\entry\src\main\java\com\example\commoneventservicetest\slice\MainAbilitySlice.java

```
1    IntentAgent intentAgent=getIntentAgent();
2    request.setIntentAgent(intentAgent);
```

接下来重点编写 getIntentAgent()方法来获取 IntentAgent 对象，详细代码如程序清单 6-16 所示。

程序清单 6-16：hmos\ch06\01\CommonEventServiceTest\entry\src\main\java\com\example\commoneventservicetest\slice\MainAbilitySlice.java

```
1    private IntentAgent getIntentAgent() {
2        //指定要启动的 Ability 的 BundleName 和 AbilityName 字段
3        Intent intent = new Intent();
```

```
4      Operation operation = new Intent.OperationBuilder()
5              .withDeviceId("")
6              .withBundleName(getBundleName())
7              .withAbilityName(MainAbility.class.getName())
8              .build();
9      intent.setOperation(operation);
10     List<Intent> intentList = new ArrayList<>();
11     intentList.add(intent);
12     //指定启动一个有页面的 Ability
13     IntentAgentInfo paramsInfo=new IntentAgentInfo(200,
   IntentAgentConstant.OperationType.START_ABILITY, IntentAgentConstant.
   Flags.UPDATE_PRESENT_FLAG, intentList, null);
14     //获取 IntentAgent 实例
15     IntentAgent agent = IntentAgentHelper.getIntentAgent(this,
   paramsInfo);
16     return agent;
17 }
```

相关代码都编好后,再次运行项目,单击发布的通知,然后返回手机主界面,下滑通知栏,单击刚刚发布的自定义通知,就会跳转到指定好的 MainAbility 的相关页面,如图 6-10 所示。

(a) 可单击通知发布效果　　(b) 单击通知跳转后的效果

图 6-10　通知跳转

6.3 日志

之前我们多次用到日志打印来演示项目,本节系统讲解一下 HarmonyOS 提供的 HiLog 日志系统。HiLog 日志系统让应用可以按照指定类型、指定级别、指定格式字符串输出日志内容,帮助开发者了解应用的运行状态,更好地调试程序。

输出日志的接口由 HiLog 类提供。在输出日志前,需要先调用 HiLog 的辅助类 HiLogLabel(int type, int domain, String tag)定义日志标签,其中参数 type 用于指定输出日志的类型,参数 domain 用于指定输出日志所对应的业务领域,参数 tag 用于指定日志标识。使用示例代码如程序清单 6-17 所示。

程序清单 6-17:hmos\ch06\01\HiLogTest\entry\src\main\java\com\example\hilogtest\slice\MainAbilitySlice.java

```
static final HiLogLabel LABEL_LOG = new HiLogLabel(HiLog.LOG_APP,
0x00201, "MY_TAG");
```

HiLog 中定义了 DEBUG、INFO、WARN、ERROR、FATAL 5 种日志级别(如程序清单 6-18 所示),并提供了对应的 debug()、info()、warn()、error()和 fatal()方法用于输出不同级别的日志。

程序清单 6-18:hmos\ch06\01\HiLogTest\entry\src\main\java\com\example\hilogtest\slice\MainAbilitySlice.java

```
1  HiLog.debug(LABEL_LOG, "这是一个 DEBUG 级别的日志");
2  HiLog.info(LABEL_LOG, "这是一个 INFO 级别的日志");
3  HiLog.warn(LABEL_LOG, "这是一个 WARN 级别的日志");
4  HiLog.error(LABEL_LOG, "这是一个 ERROR 级别的日志");
5  HiLog.fatal(LABEL_LOG, "这是一个 FATAL 级别的日志")
```

当项目运行后,HiLog 选项卡里会有很多日志信息输出,这时就要对输出的日志信息进行筛选。那么,怎么筛选呢?打开 HiLog 选项卡,上面有一排选项卡供选择:第一个选项卡用来选择是哪台设备输出的日志信息;第二个选项卡用来选择是哪个应用输出的日志信息;第三个选项卡用来筛选某一种级别的日志信息;第四个选项卡用来筛选某指定日志标识的日志信息,如图 6-11 所示。

图 6-11 日志信息输出台

6.4 本章小结

本章介绍了公共事件和通知的开发方式以及日志的使用方法。HarmonyOS 中的公共事件和 Android 中的广播的作用类似，其本质上是一种全局的监听器，发布公共事件需要借助 CommonEventData 对象，接收公共事件需要继承 CommonEventSubscriber 类并实现 onReceiveEvent 回调函数。本章详细介绍了四种公共事件的使用方法，此外还介绍了通知的基本使用方法，通过使用 IntentAgent 达到单击通知栏会跳转到 App 内指定界面的效果。最后，本章还介绍了在应用开发阶段经常用来作调试作用的日志功能。

6.5 课后习题

1. 公共事件有几种类型？分别是什么？
2. 有序公共事件有什么特点？
3. 简述通知开发的大致步骤。
4. 日志有几种级别？分别是什么？

第 7 章

综合案例——"远程闹钟"

课程思政 7

本章要点

- "远程闹钟"案例需求概述
- "远程闹钟"程序结构
- 界面设计
- 服务器搭建与部署

本章知识结构图（见图 7-1）

图 7-1　本章知识结构图

本章示例（见图 7-2）

图 7-2　本章示例图

通过前面章节的学习，我们已经掌握了 HarmonyOS 的基础知识，本章通过一个带服务器端设计与开发的综合项目案例将前面所学的知识串联起来，共同开发一个实用的应用程序。本章将带领读者从零开始开发一个"远程闹钟"应用程序，具体分为手机端应用和手表端应用。该应用主要是为了解决家长不在家时无法按时叫醒家里小孩的问题，共包括闹钟查询、闹钟添加、闹钟删除、闹铃播放 4 个模块。

在手机端应用程序中，闹钟查询模块作为主界面，用户可以直接在此界面查看已经设置好的闹钟的当前情况。用户单击应用右上角的按钮时，将跳转到闹钟添加模块，用户可在此界面设置新的闹钟，单击该界面右上角的按钮就可保存并返回到手机端应用主界面；单击左上角的按钮则不进行新闹钟的设置，而是直接返回到手机端应用的主界面。当闹钟列表里有闹钟信息时，用户可以通过单击任意一个闹钟信息栏进入闹钟删除模块，在该界面中用户可以选择需要删除的闹钟。当某一闹钟已发生过时，手机端主界面的闹钟列表中的相应项目栏的右侧会显示图标用来表示该闹钟已发生。

在手表端应用程序中，当有闹铃需要发生时，应用程序将会跳转到闹铃播放模块，当闹铃播放完成后，应用程序会自动跳转到手表端应用主界面。当闹铃正在播放时，用户也可以单击"关闭"按钮提前关闭闹铃，返回手表端主界面。

本章案例所涉及的知识包括：基本界面控件的使用，如 Text、Button、Image、Checkbox、TimePicker、DatePicker 等；高级界面控件的使用，如 ListContainer、CommonDialog 等；Ability Slice 之间的跳转；事件处理，如单击事件、选择事件等；以及关系数据库的使用。

通过本章的学习，读者对这些知识会有更深入的了解，能够自主开发一些小应用。

7.1 "远程闹钟"概述

"远程闹钟"应用软件包含手机端应用和手表端应用。手机端应用主要为家长提供给孩子设置闹钟的功能，包含闹钟查询、闹钟添加、闹钟删除3个具体功能。手表端应用则为闹铃播放功能，当手机端设置的闹钟时间到了，闹铃将自动响起。对于手机端和手表端之间的信息传输，我们使用 Spring Boot 搭建一个后台服务器进行数据传输，其功能结构图如图 7-3 所示，程序结构图如图 7-4 所示。

图 7-3 "远程闹钟"功能结构图

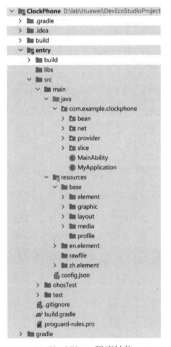

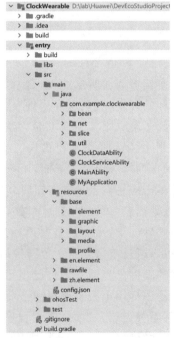

(a) ClockPhone程序结构　　(b) ClockWearable程序结构　　(c) hmostest程序结构

图 7-4 "远程闹钟"项目程序结构图

7.2 Spring Boot 服务器端设计

7.2.1 Spring Boot 技术简介

Spring Boot 是由 Pivotal 团队提供的全新框架,其设计目的是简化 Spring 应用的初始搭建以及开发过程。该框架使用了特定的方式进行配置,从而使开发人员不再需要定义样板化的配置。Spring Boot 其实就是一个整合了很多可插拔组件(框架),内嵌了使用工具(如 Tomcat、Jetty 等),方便开发人员快速搭建和开发的一个框架。

7.2.2 Spring Boot 项目开发环境

1. Maven 项目管理工具的安装与配置

Maven 是项目构建和依赖管理工具,通过 pom.xml 文件中的 dependency 属性管理依赖的 jar 包。Maven 把所有的 Java 代码都放在 src/main/java 下,把所有的测试代码都放在 src/test/java 下,这样便方便了开发者。

(1) Maven 的下载和安装。

从 Maven 官网(https://maven.apache.org/download.cgi)下载 Apache Maven 3.8.1,其中 Windows 操作系统的用户选择 apache-maven-3.8.1-bin.zip;Linux 和 macOS 操作系统的用户选择 apache-maven-3.8.1-bin.tar.gz,如图 7-5 所示。

Binary tar.gz archive	apache-maven-3.8.1-bin.tar.gz
Binary zip archive	apache-maven-3.8.1-bin.zip
Source tar.gz archive	apache-maven-3.8.1-src.tar.gz
Source zip archive	apache-maven-3.8.1-src.zip

图 7-5 Maven 的下载

(2) 配置 Maven 环境变量。

Maven 下载并安装成功之后,右击"计算机/此电脑"图标,选择"属性"选项,之后单击"高级系统设置">"环境变量"设置环境变量。编辑系统变量 Path,添加变量值 D:\lab\Maven\apache-maven-3.8.1\bin(该路径根据用户解压 Maven 压缩包的路径所决定),如图 7-6 所示。

2. Spring Boot 项目搭建

首先打开 IntelliJ IDEA 编译器,单击 Creat New Project 创建项目,选择 Spring Initializr,输入 Name,设置 Location 与相应的 JDK 版本后单击 Next 按钮,如图 7-7 所示。

单击 Next 按钮后,选择需要添加的依赖,此处选择 Spring Web,之后单击 Finish 按钮,等待依赖下载完成,Spring Boot 项目便创建完成,如图 7-8 所示。

图 7-6　Maven 环境变量的配置

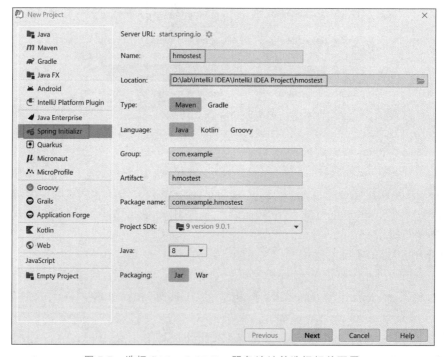

图 7-7　选择 Spring Initializr 服务地址并选择相关配置

第 7 章 综合案例——"远程闹钟"

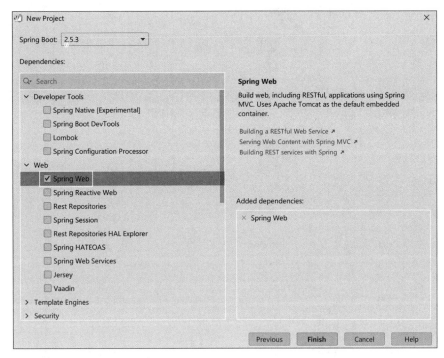

图 7-8　Spring Web 项目配置

7.2.3　数据库设计

在该项目中涉及闹钟时间以及闹钟状态等数据，所以这里设计一张闹钟（clock）表记录这些数据，该表的结构如表 7-1 所示。

表 7-1　clock 表结构

编号	名　称	字段代码	类型	长度	是否可空	主键
1	闹钟编号	clockid	int	255	非空	是
2	闹钟年份	years	int	255	非空	否
3	闹钟月份	months	int	255	非空	否
4	闹钟日期	days	int	255	非空	否
5	闹钟小时	hours	int	255	非空	否
6	闹钟分钟	minutes	int	255	非空	否
7	是否送达到手表	sends	int	255	可空	否
8	闹钟是否已响	happened	int	255	可空	否

设计好表的结构后，就把这些设计好的结构建立起来。打开 Navicat for MySQL 软件，单击"连接"按钮（图 7-9），进入"新建连接"界面，"连接名"这一栏可以随意命名，这里输入 hmostest，"密码"一栏是安装 MySQL 时自己设定的密码（图 7-10），之后单击"确定"按钮。

图 7-9　新建连接

图 7-10　新建连接相关信息

输入完连接名和密码之后,单击"确定"按钮,可以发现连接图标下方出现了名为 hmostest 的新连接,然后双击 hmostest 新连接,右键选择"新建数据库"选项(图 7-11)。这里数据库名也写为 hmostest,字符集选择 utf8,排序规则选择 utf8_general_ci,然后单击"确定"按钮即可(图 7-12)。

图 7-11　新建数据库

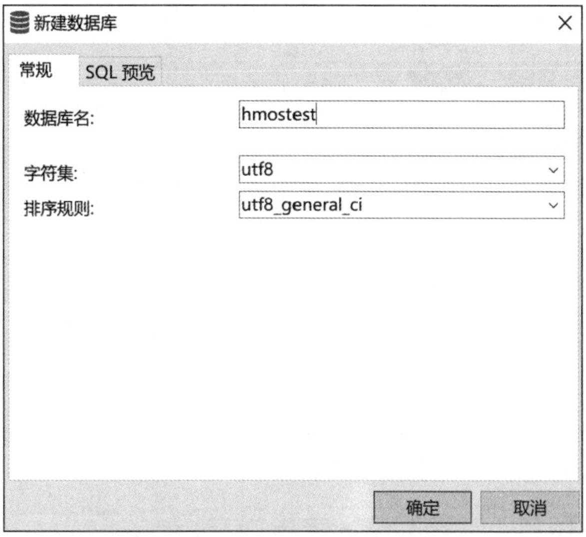

图 7-12 新建数据库相关信息

新建数据库后,要在新建的数据库中新建需要的数据表。选择 hmostest 数据库下的"表"选项并右击,从弹出的快捷菜单中选择"新建表"选项,如图 7-13 所示。

图 7-13 新建表

之后右边出现新建表的界面,根据数据表的设计在新建的表中添加"名""类型""长度"和"不是 null"等信息。按照图 7-14 中的示例填写好后,单击左上角的"保存"按钮保存此表并把它命名为 clock。

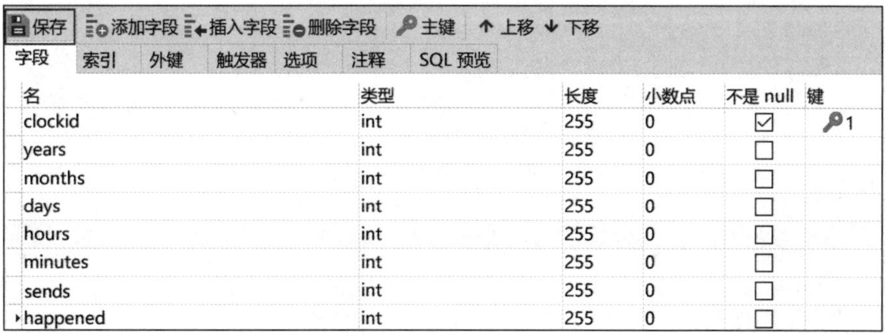

名	类型	长度	小数点	不是 null	键
clockid	int	255	0	☑	🔑1
years	int	255	0	☐	
months	int	255	0	☐	
days	int	255	0	☐	
hours	int	255	0	☐	
minutes	int	255	0	☐	
sends	int	255	0	☐	
happened	int	255	0	☐	

图 7-14 新建表信息

保存后,表的效果如图 7-15 所示。

图 7-15　新建表效果图

7.2.4 "远程闹钟"服务器搭建

1. 通过配置文件实现数据库的连接配置

打开项目目录,找到 application.properties 文件。此处,为了方便后面的项目配置,右击该文件,从弹出的快捷菜单中选择 Refactor 选项,之后再选择 Rename 选项,将该文件重命名为 application.yml,如图 7-16 所示。

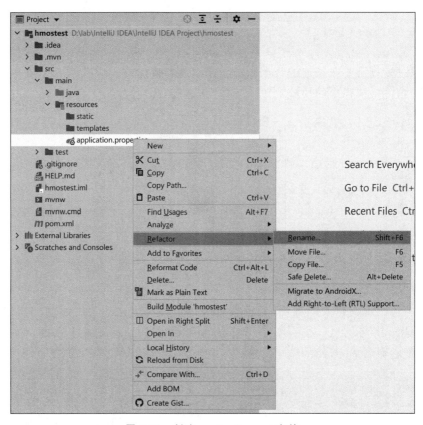

图 7-16　创建 application.yml 文件

打开 application.yml 文件,按照程序清单 7-1 所示输入代码。

程序清单 7-1：hmos\ch07\01\hmostest\src\main\resources\application.yml

```
1  spring:
```

```
2    datasource:
3      jdbc:mysql://localhost:3306/hmostest?useUnicode=
  true&characterEncoding=UTF-8&useSSL=false
4      username: root
5      password:
6      driver-class-name: com.mysql.jdbc.Driver
```

url 代表数据库的连接地址；username 为数据库用户名，password 为密码；driver-class-name 为驱动名称。

注意，此时驱动名称之所以会标红，原因是没有在 Maven 的 pom.xml 配置文件中添加 mysql 数据库依赖。这时需要打开项目目录，之后打开 pom.xml 配置文件，向其中添加如程序清单 7-2 所示的依赖。

程序清单 7-2：hmos\ch07\01\hmostest\pom.xml

```
1  <dependency>
2      <groupId>mysql</groupId>
3      <artifactId>mysql-connector-java</artifactId>
4      <scope>runtime</scope>
5  </dependency>
```

2. 使用 mybatis 对数据库进行管理

（1）导入依赖关系并配置 mybatis 文件。找到项目目录的 pom.xml 文件，单击进入，添加如程序清单 7-3 所示的依赖。

程序清单 7-3：hmos\ch07\01\hmostest\pom.xml

```
1  <dependency>
2      <groupId>log4j</groupId>
3      <artifactId>log4j</artifactId>
4      <version>1.2.17</version>
5  </dependency>
6  <dependency>
7      <groupId>org.mybatis.spring.boot</groupId>
8      <artifactId>mybatis-spring-boot-starter</artifactId>
9      <version>2.1.4</version>
10 </dependency>
```

之后，进入 application.yml 配置文件，添加 mybatis 的映射与实体路径的配置语句，如程序清单 7-4 所示。

程序清单 7-4：hmos\ch07\01\hmostest\src\main\resources\application.yml

```
1  mybatis:
```

```
2       mapper-locations: classpath:mapper/*Mapper.xml
3       type-aliases-package: com.example.demo.bean
```

(2) 配置 mybatis-generator.xml 文件。进入项目目录的 src/resources 文件夹,新建文件,并命名为 generatorConfig.xml,输入如程序清单 7-5 所示的配置代码。

程序清单 7-5: hmos\ch07\01\hmostest\src\main\resources\generatorConfig.xml

```
1   <?xml version="1.0" encoding="UTF-8"?>
2   <!DOCTYPE generatorConfiguration
3       PUBLIC "-//mybatis.org//DTD MyBatis Generator Configuration 1.0//EN"
4       "http://mybatis.org/dtd/mybatis-generator-config_1_0.dtd">
5   <generatorConfiguration>
6     <classPathEntry
7         location="D:\lab\MavenRepository\mysql-connector-java-5.1.9.jar"/>
8     <context id="context1" targetRuntime="MyBatis3">
9       <commentGenerator>
10          <!-- 去除自动生成的注释 -->
11          <property name="suppressAllComments" value="true"/>
12      </commentGenerator>
13      <!-- 数据库连接配置 -->
14      <jdbcConnection driverClass="com.mysql.jdbc.Driver"
15              connectionURL="jdbc:mysql://localhost:3306/hmostest"
16              userId="root"
17              password=""/>
18      <!-- 非必需,类型处理器,在数据库类型和 Java 类型之间的转换控制-->
19      <javaTypeResolver>
20          <property name="forceBigDecimals" value="false"/>
21      </javaTypeResolver>
22      <!--配置生成的实体包
23          targetPackage:生成的实体包位置,默认存放在 src 目录下
24          targetProject:目标工程名
25      -->
26      <javaModelGenerator targetPackage="com.example.hmostest.bean"
27              targetProject="src/main/java"/>
28      <!-- 实体包对应映射文件位置及名称,默认存放在 src 目录下 -->
29      <sqlMapGenerator targetPackage="mapper"
30              targetProject="src/main/resources"/>
31      <javaClientGenerator targetPackage="com.example.hmostest.dao"
32              targetProject="src/main/java" type="XMLMAPPER"/>
33      <table schema="" tableName="clock" enableCountByExample="false" enableSelectByExample="false"
```

```
34              enableDeleteByExample="false" enableUpdateByExample=
   "false" selectByExampleQueryId="false">
35          </table>
36      </context>
37  </generatorConfiguration>
```

（3）逆向建立项目的 bean、dao、mapper 的文件。首先进入 pom.xml 文件，添加如程序清单 7-6 所示的依赖。

程序清单 7-6：hmos\ch07\01\hmostest\pom.xml

```
1  <dependency>
2      <groupId>org.mybatis.generator</groupId>
3      <artifactId>mybatis-generator-core</artifactId>
4      <version>1.3.6</version>
5  </dependency>
```

然后在＜build＞标签内添加如程序清单 7-7 所示的插件。

程序清单 7-7：hmos\ch07\01\hmostest\pom.xml

```
1  <plugin>
2      <groupId>org.mybatis.generator</groupId>
3      <artifactId>mybatis-generator-maven-plugin</artifactId>
4      <version>1.3.6</version>
5      <!--在 pom.xml 中配置 MBG 插件时，可以通过 configuration 标签指定 MBG 的配
   置文件名、是否覆盖同名文件、是否将生成过程输出至控制台等-->
6      <configuration>
7          <!--MyBaits-generator 的配置文件 generatorconfig.xml 的位置-->
8          <configurationFile>src/main/resources/generatorConfig.xml</
   configurationFile>
9          <!--是否将生成过程输出至控制台-->
10         <verbose>true</verbose>
11         <!--是否覆盖同名文件(只是针对 XML 文件，Java 文件生成类似＊.java.1、＊.
   java.2 形式的文件)-->
12         <overwrite>true</overwrite>
13     </configuration>
14 <plugin>
```

最后，如图 7-17 所示，单击右上角的 Edit Configurations 选项，进入选择界面。选择左上角的"＋"按钮，之后选择添加 Maven 项目，如图 7-18 所示。

如图 7-19 所示，在 Command line 内输入 mybatis-generator：generate -e。

如图 7-20 所示，单击 RUN 按钮后运行该 Maven 项目。

打开项目目录，如果出现如图 7-21 所示的包与类，则表示工程建立成功。

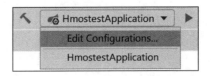

图 7-17　单击 Edit Configurations 选项

图 7-18　选择添加 Maven 项目

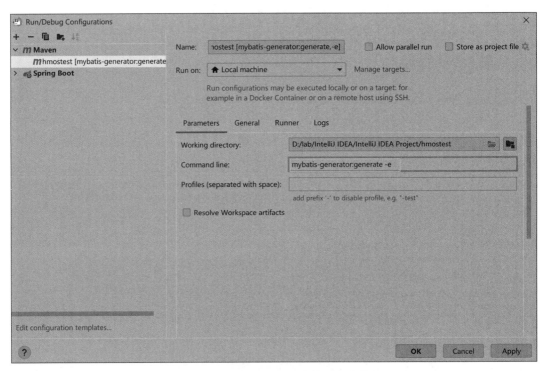

图 7-19　在 Command line 内输入相应信息

图 7-20　启动 Maven 项目

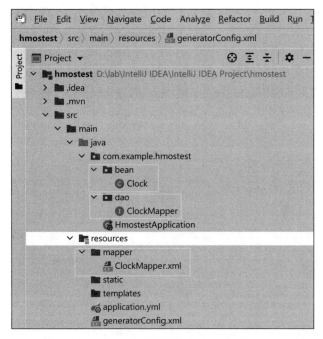

图 7-21　Maven 项目启动成功的项目目录

3. 数据库基本操作的实现

接下来,使用上面通过构造器创建的几个文件提供的功能实现数据库基本操作。

右击 com.example.hmostest 目录,选择 New 选项,然后单击 Package,并命名为 controller。右击 controller 包,选择 New 选项,然后单击 Java Class,将文件命名为 ClockController,如图 7-22 所示。

之后,在 ClockController 类中添加注解@RestController,告诉 Spring Boot 此类作为控制器。由于要使用 ClockMapper 接口,因此首先初始化 AreaMapper 接口,并且通过注解@Autowired 进行标记,详细代码如程序清单 7-8 所示。

图 7-22　新建 ClockController

程序清单 7-8:hmos\ch07\01\hmostest\src\main\java\com\example\hmostest\controller\ClockController.java

```
1    @RestController
2    public class ClockController {
3    
4        @Autowired
5        private ClockMapper clockMapper;
6    }
```

然后创建一个 SetNewClock()方法,用来增加新的闹钟信息到数据库中,并且在方法 SetNewClock()上方添加注解@RequestMapping,告诉 Servlet 我们的请求路径,其中路径名为 @RequestMapping 注解括号的内容,此处以/setNewClock 为路径名。与 SetNewClock()方法类似,再创建 4 个方法:SetSends()用于设置 send 值为 1 表示该闹钟数据已被存储到手表端;SetHappened()用于设置 happened 值为 1 表示该闹钟已响;DeleteClock()用于删除指定 clockid 的闹钟;SearchAllClock()用于获取数据库里所有的闹钟信息,详细代码如程序清单 7-9 所示。

程序清单 7-9: hmos\ch07\01\hmostest\src\main\java\com\example\hmostest\controller\ClockController.java

```
1    @RequestMapping("/setNewClock")
2    public void SetNewClock(@RequestParam(value = "clockid", defaultValue = " ") int clockid,
3                            @RequestParam(value = "years", defaultValue = " ") int years,
4                            @RequestParam(value = "months", defaultValue = "") int months,
5                            @RequestParam(value = "days", defaultValue = "") int days,
6                            @RequestParam(value = "hours", defaultValue = "") int hours,
7                            @RequestParam(value = "minutes", defaultValue = "") int minutes,
8                            @RequestParam(value = "sends", defaultValue = "0") int sends,
9                            @RequestParam(value = "happened", defaultValue = "0") int happened) {
10       Clock clock = new Clock();
11       clock.setClockid(clockid);
12       clock.setYears(years);
13       clock.setMonths(months);
14       clock.setDays(days);
15       clock.setHours(hours);
16       clock.setMinutes(minutes);
17       clock.setSends(sends);
18       clock.setHappened(happened);
19       clockMapper.insert(clock);
20    }
21
22    @RequestMapping("/setSends")
23    public void SetSends(@RequestParam(value = "clockid", defaultValue = " ") int clockid) {
```

```
24        Clock clock = clockMapper.selectByPrimaryKey(clockid);
25        clock.setSends(1);
26        clockMapper.updateByPrimaryKey(clock);
27    }
28
29    @RequestMapping("/setHappened")
30    public void SetHappened(@RequestParam(value = "clockid", defaultValue = " ")
      int clockid) {
31        Clock clock = clockMapper.selectByPrimaryKey(clockid);
32        clock.setHappened(1);
33        clockMapper.updateByPrimaryKey(clock);
34    }
35
36    @RequestMapping("/deleteClock")
37    public void DeleteClock(@RequestParam(value = "clockid", defaultValue = " ")
      int clockid) {
38        clockMapper.deleteByPrimaryKey(clockid);
39    }
40
41    @RequestMapping("/searchAllClock")
42    public List<Clock> SearchAllClock() {
43        return clockMapper.selectAll();
44    }
```

编写好上面的代码后,会发现 SearchAllClock() 方法中的 selectAll() 标红报错,这是因为在 ClockMapper 中并没有 selectAll() 这个方法,既然如此,那我们就创建一个 selectAll() 方法(见程序清单 7-10)。

程序清单 7-10: hmos\ch07\01\hmostest\src\main\java\com\example\hmostest\dao\ClockMapper.java

```
List<Clock> selectAll();
```

然后,在 resources/mapper 文件夹下的 ClockMapper.xml 文件里的"mapper"字段下添加如程序清单 7-11 所示的代码。

程序清单 7-11: hmos\ch07\01\hmostest\src\main\resources\mapper\ClockMapper.xml

```
1  <select id="selectAll" resultMap="BaseResultMap">
2      select *
3      from clock
4      order by clockid asc
5  </select>
```

最后,在 com/example/hmostest 下的 HmostestApplication 中添加一个注释,详细代码如程序清单 7-12 所示。

程序清单 7-12 hmos\ch07\[E]\hmostest\src\main\java\com\example\hmostest\HmostestApplication.java

```
1   @SpringBootApplication
2   @MapperScan("com.example.hmostest.dao")
3   public class HmostestApplication {
4
5       public static void main(String[] args) {
6           SpringApplication.run(HmostestApplication.class, args);
7       }
8
9   }
```

至此,服务器的代码部分已经全部编写完毕。

4. 项目运行测试

选择 HmostestApplication 进行运行,当控制台显示如图 7-23 所示的日志输出时,表示项目启动成功。

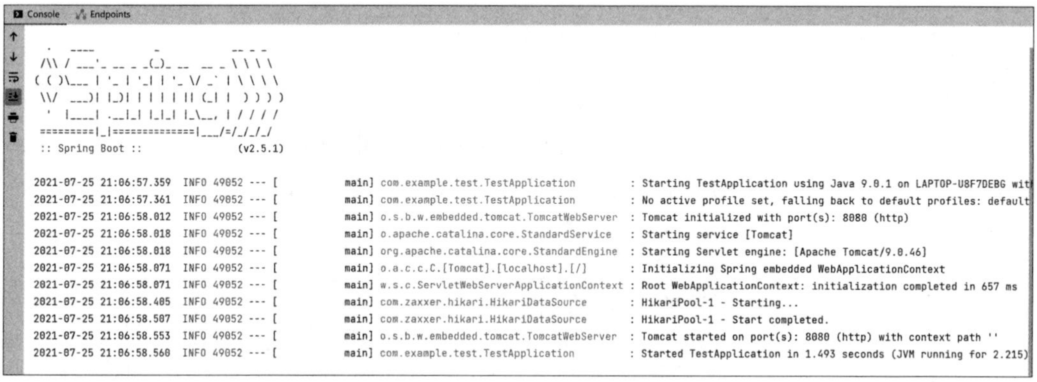

图 7-23　Spring Boot 项目启动成功效果图

项目启动成功后,测试一下看项目是否运行成功。

(1) 插入数据。

插入一条 clockid 为 2107252112,时间为 2021 年 7 月 25 日 21 时 12 分的闹钟数据。打开浏览器,输入

```
http://localhost:8080/setNewClock?clockid=2107252112&years=2021&months=7&days=25&hours=21&&minutes=12
```

若数据库中出现如图 7-24 所示的数据,则表示项目运行成功。

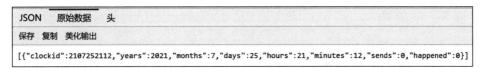

图 7-24 闹钟信息添加成功

(2) 查询所有闹钟信息。

查询数据库里所有的闹钟信息,此时数据库中只有刚刚插入的 clockid 为 2107252112 的闹钟数据。

打开浏览器,输入

```
http://localhost:8080/searchAllClock,
```

若出现如图 7-25 所示的数据,则表示项目运行成功。

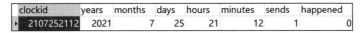

图 7-25 数据查询成功

(3) 设置 sends 的值。

当闹钟信息被手表端获取后,将 sends 的值设置为 1。

打开浏览器,输入

```
http://localhost:8080/setSends?clockid=2107252112
```

如图 7-26 所示,若数据库中 clockid 为 2107252112 的闹钟的 sends 值变为 1,则表示项目运行成功。

图 7-26 设置 sends 值成功

(4) 设置 happened 的值。

当闹钟在手表端按时响起后,将 happened 的值设置为 1。

打开浏览器,输入

```
http://localhost:8080/setHappened?clockid=2107252112
```

如图 7-27 所示,若数据库中 clockid 为 2107252112 的闹钟的 happened 值变为 1,则表示项目运行成功。

clockid	years	months	days	hours	minutes	sends	happened
2107252112	2021	7	25	21	12	1	1

图 7-27 设置 happened 值成功

(5) 删除指定闹钟。

删除数据库中 clockid 为 2107252112 的闹钟。

打开浏览器,输入

```
http://localhost:8080/deleteClock?clockid=2107252112
```

如图 7-28 所示,若数据库中 clockid 为 2107252112 的闹钟被删除,则表示项目运行成功。

图 7-28　删除闹钟信息成功

7.2.5　部署服务器

因为目前 HUAWEI DevEco Studio 提供的虚拟机为远程虚拟机,并且目前国内运营商因为各种原因分配给用户的大都为内网 IP,因此需要把服务器部署到公网 IP 上。关于公网 IP,有多种获取方式,最常见的为向一些商家租赁一个服务器,例如阿里云、百度云等,但是这种方式的费用较高,最普通的服务器一年的租赁费都在几百元。学习者通常对服务器的访问速度没有过于硬性的要求,所以可以使用内网穿透。内网穿透也叫作 NAT 穿透,它的实质是利用路由器上的 NAT 系统将私有(保留)地址转化为合法 IP 地址的转换技术。那么,如何简易地实现内网穿透呢? 这里使用钉钉开放平台提供的工具实现内网穿透。进入钉钉开放平台首页(https://developers.dingtalk.com/),在网页中搜索"内网穿透"就能找到钉钉开放平台给开发者提供的内网穿透的方法,如图 7-29 所示。

图 7-29　钉钉开放平台的内网穿透工具

进入钉钉开放平台提供的 GitHub 项目网址下载内网穿透工具,之后进入 CMD 命令模式,切换到工具所在目录后输入命令 ding -config=ding.cfg -subdomain=abcdd 8080,执行该命令后,若出现如图 7-30 所示的窗口,则表示启动内网穿透成功。

此时,打开浏览器,输入

```
http://abcdd.vaiwan.com/setNewClock?clockid=2107252112&years=2021&months=7&days=25&hours=21&&minutes=12
```

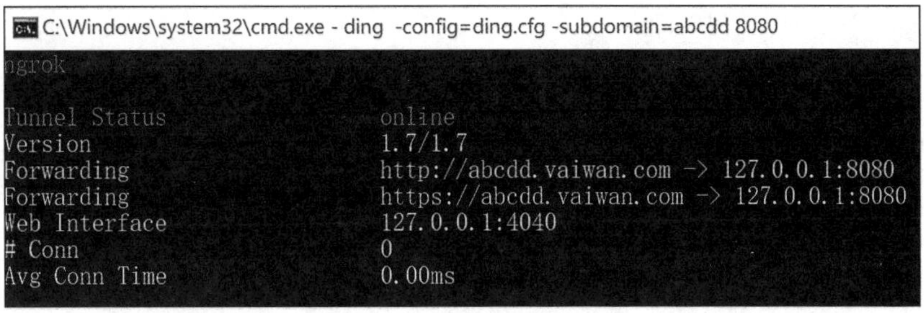

图 7-30　内网穿透成功

若数据库中出现如图 7-31 所示的数据，则表示内网穿透成功。

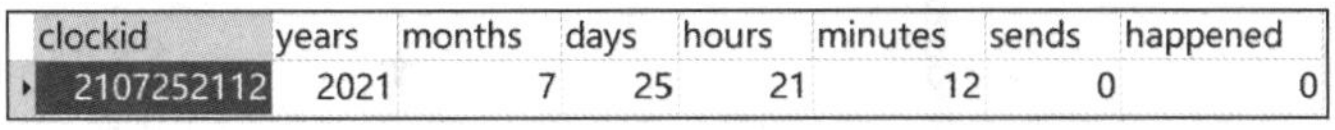

图 7-31　闹钟信息添加成功

7.3　"远程闹钟"手机端应用设计

服务器端应用写好后，下面来写手机端应用。手机端应用主要实现的功能是闹钟的设置和删除，使用者可以在手机上设置好闹钟后传到后台服务器，也可以选择性地删除闹钟。手机端软件主要包括闹钟显示、闹钟添加、闹钟删除 3 个模块。新建一个项目，命名为 ClockPhone，如图 7-32 所示。

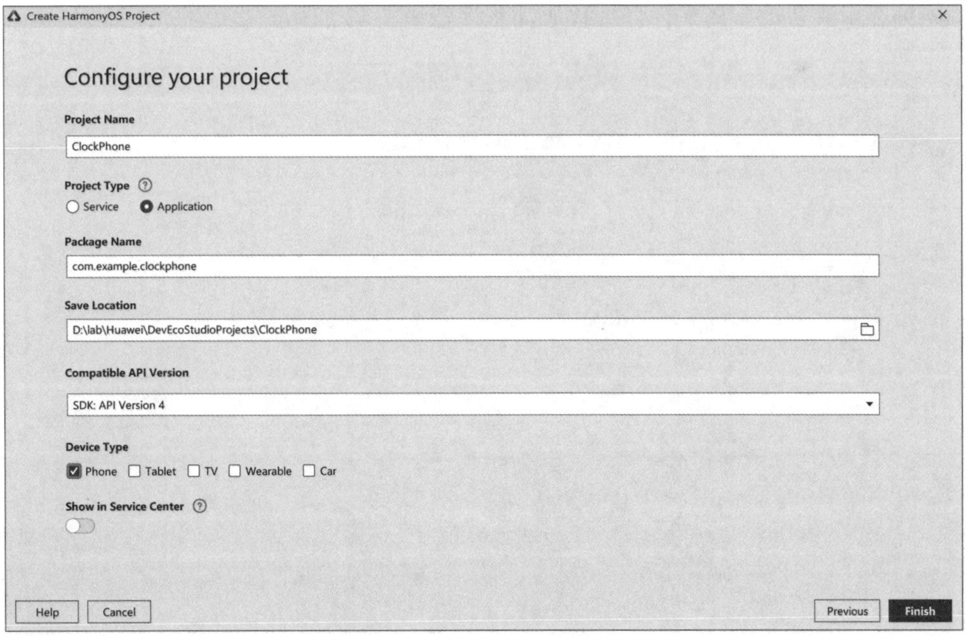

图 7-32　创建 ClockPhone 项目

7.3.1 闹钟显示模块

闹钟显示模块的界面是软件的主界面，其中包含一个 ListContainer，用于显示闹钟列表，其预览效果如图 7-33 所示。

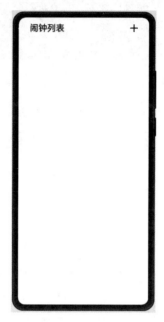

图 7-33　闹钟显示模块界面预览效果

闹钟显示模块的布局文件详细代码如程序清单 7-13 所示。

程序清单 7-13：hmos\ch07\01\ClockPhone\entry\src\main\resources\base\layout\ability_main.xml

```
1   <?xml version="1.0" encoding="utf-8"?>
2   <DirectionalLayout
3       xmlns:ohos="http://schemas.huawei.com/res/ohos"
4       ohos:height="match_parent"
5       ohos:width="match_parent"
6       ohos:alignment="center"
7       ohos:orientation="vertical">
8       <DirectionalLayout
9           ohos:id="$+id:layout_clock"
10          ohos:height="match_parent"
11          ohos:width="match_parent"
12          ohos:orientation="vertical">
13          <DependentLayout
14              ohos:height="60vp"
15              ohos:width="match_parent"
16              ohos:padding="10vp">
```

```
17            <Text
18                ohos:id="$+id:text_clockname"
19                ohos:height="match_content"
20                ohos:width="match_content"
21                ohos:align_parent_left="true"
22                ohos:background_element="$graphic:background_ability_main"
23                ohos:layout_alignment="horizontal_center"
24                ohos:left_margin="20vp"
25                ohos:text="闹钟列表"
26                ohos:text_size="25fp"
27                ohos:vertical_center="true"/>
28            <Image
29                ohos:id="$+id:image_add_clock"
30                ohos:height="25vp"
31                ohos:width="25vp"
32                ohos:align_parent_right="true"
33                ohos:foreground_element="$media:add"
34                ohos:right_margin="20vp"
35                ohos:vertical_center="true"/>
36        </DependentLayout>
37        <ListContainer
38            ohos:id="$+id:list_clock_view"
39            ohos:height="match_content"
40            ohos:width="match_parent"/>
41    </DirectionalLayout>
42 </DirectionalLayout>
```

ListContainer 中的每条闹钟信息的布局文件详细代码如程序清单 7-14 所示。

程序清单 7-14：hmos\ch07\01\ClockPhone\entry\src\main\resources\base\layout\item_clock.xml

```
1  <?xml version="1.0" encoding="utf-8"?>
2  <DependentLayout
3      xmlns:ohos="http://schemas.huawei.com/res/ohos"
4      ohos:height="match_parent"
5      ohos:width="match_parent"
6      ohos:orientation="vertical">
7      <Text
8          ohos:id="$+id:item_text_date"
9          ohos:height="match_content"
10         ohos:width="match_content"
11         ohos:layout_alignment="center"
12         ohos:left_margin="20vp"
13         ohos:padding="4vp"
```

```
14          ohos:text="Date"
15          ohos:text_size="20fp"/>
16      <Text
17          ohos:id="$+id:item_text_time"
18          ohos:height="match_content"
19          ohos:width="match_content"
20          ohos:below="$id:item_text_date"
21          ohos:layout_alignment="center"
22          ohos:left_margin="20vp"
23          ohos:padding="4vp"
24          ohos:text="Time"
25          ohos:text_size="20fp"/>
26      <Component
27          ohos:id="$+id:item_component_state"
28          ohos:height="30vp"
29          ohos:width="60vp"
30          ohos:align_parent_right="true"
31          ohos:background_element="$media:yes"
32          ohos:right_margin="20vp"
33          ohos:top_margin="20vp"/>
34      <Component
35          ohos:height="1vp"
36          ohos:width="match_parent"
37          ohos:background_element="$graphic:list_divider"
38          ohos:below="$id:item_text_time"/>
39  </DependentLayout>
```

这里需要给 ListContainer 写一个适配器，控制数据的显示。新建一个 provider 包，在该包下新建一个 ClockListItemProvider 类，该类继承 BaseItemProvider，详细代码如程序清单 7-15 所示。

程序清单 7-15：hmos\ch07\01\ClockPhone\entry\src\main\java\com\example\clockphone\provider\ClockListItemProvider.java

```
1  public class ClockListItemProvider extends BaseItemProvider {
2      private AbilitySlice slice;
3      private List<Clock> dataList = new LinkedList<>();
4      private SimpleDateFormat simpleDateFormat1 = new SimpleDateFormat
   ("HH:mm");
5      private SimpleDateFormat simpleDateFormat2 = new SimpleDateFormat
   ("yyyy-MM-dd");
6
7      public ClockListItemProvider(AbilitySlice abilitySlice) {
```

```
8            this.slice = abilitySlice;
9        }
10
11       @Override
12       public int getCount() {
13           return dataList.size();
14       }
15
16       @Override
17       public Object getItem(int i) {
18           return dataList.get(i);
19       }
20
21       @Override
22       public long getItemId(int i) {
23           return i;
24       }
25
26       public void setDataList(List<Clock> dataList) {
27           this.dataList = dataList;
28       }
29
30       @Override
31       public Component getComponent(int position, Component component,
   ComponentContainer componentContainer) {
32           Component component1;
33           ViewHolder viewHolder;
34           if (component == null) {
35               component1 = LayoutScatter.getInstance(slice).parse
   (ResourceTable.Layout_item_clock, null, false);
36               viewHolder = new ViewHolder();
37               viewHolder.textDate = (Text) component1.findComponentById
   (ResourceTable.Id_item_text_date);
38               viewHolder.textTime = (Text) component1.findComponentById
   (ResourceTable.Id_item_text_time);
39               viewHolder.component_clock_state = component1.
   findComponentById(ResourceTable.Id_item_component_state);
40               component1.setTag(viewHolder);
41           } else {
42               component1 = component;
43               viewHolder = (ViewHolder) component.getTag();
44           }
45           Clock clock = dataList.get(position);
```

```
46          Calendar calendar = Calendar.getInstance();
47          calendar.set(Calendar.YEAR, clock.getYears());
48          calendar.set(Calendar.MONTH, clock.getMonths() - 1);
49          calendar.set(Calendar.DAY_OF_MONTH, clock.getDays());
50          calendar.set(Calendar.HOUR_OF_DAY, clock.getHours());
51          calendar.set(Calendar.MINUTE, clock.getMinutes());
52          viewHolder.textTime.setText(simpleDateFormat1.format(calendar.
    getTime()));
53          viewHolder.textDate.setText(simpleDateFormat2.format(calendar.
    getTime()));
54          calendar.clear();
55          if (clock.getHappened() == 0) {//闹钟没有响完时不显示图标
56              viewHolder.component_clock_state.setBackground(null);
57          }
58          return component1;
59      }
60
61      class ViewHolder {
62          Text textDate;
63          Text textTime;
64          Component component_clock_state;
65      }
66  }
```

此时可以注意到有关 Clock 的代码均标红报错,这是因为还没有写 Clock 的 JavaBean。现在新建一个 bean 包,在该包下新建一个 Java 类,并命名为 Clock,详细代码如程序清单 7-16 所示。

程序清单 7-16:hmos\ch07\01\ClockPhone\entry\src\main\java\com\example\clockphone\bean\Clock.java

```
1   public class Clock {
2       private Integer clockid;
3       private Integer years;
4       private Integer months;
5       private Integer days;
6       private Integer hours;
7       private Integer minutes;
8       private Integer sends;
9       private Integer happened;
10
11      public Integer getClockid() {
12          return clockid;
```

```java
13      }
14
15      public void setClockid(Integer clockid) {
16          this.clockid = clockid;
17      }
18
19      public Integer getYears() {
20          return years;
21      }
22
23      public void setYears(Integer years) {
24          this.years = years;
25      }
26
27      public Integer getMonths() {
28          return months;
29      }
30
31      public void setMonths(Integer months) {
32          this.months = months;
33      }
34
35      public Integer getDays() {
36          return days;
37      }
38
39      public void setDays(Integer days) {
40          this.days = days;
41      }
42
43      public Integer getHours() {
44          return hours;
45      }
46
47      public void setHours(Integer hours) {
48          this.hours = hours;
49      }
50
51      public Integer getMinutes() {
52          return minutes;
53      }
54
55      public void setMinutes(Integer minutes) {
```

```
56          this.minutes = minutes;
57      }
58
59      public Integer getSends() {
60          return sends;
61      }
62
63      public void setSends(Integer sends) {
64          this.sends = sends;
65      }
66
67      public Integer getHappened() {
68          return happened;
69      }
70
71      public void setHappened(Integer happened) {
72          this.happened = happened;
73      }
74  }
```

接下来,从服务器端获取数据,对服务器端传来的 JSON 数据使用 HiJson 解析。HiJson 是华为公司推荐的用于 JSON 数据解析的工具类,它的使用需要先在 entry 下的配置文件(build.gradle)中进行导入,如图 7-34 所示。

```
dependencies {
    implementation fileTree(dir: 'libs', include: ['*.jar', '*.har'])
    testImplementation 'junit:junit:4.13'
    ohosTestImplementation 'com.huawei.ohos.testkit:runner:1.0.0.100'
    implementation group: 'org.devio.hi.json', name: 'hijson', version: '1.0.0'
}
```

图 7-34 导入 HiJson

接着新建一个包 net,在该包下新建一个接口,把它命名为 NetIf,详细代码如程序清单 7-17 所示。

程序清单 7-17: hmos\ch07\01\ClockPhone\entry\src\main\java\com\example\clockphone\net\NetIf.java

```
1  public interface NetIf {
2      void setClock(Map<String, Integer> params, NetListener listener);
3
4      void getAllClock(NetListener listener);
5
6      void deleteClock(int[] clockid, NetListener listener);
```

```
7
8      interface NetListener {
9          void onSuccess(HiJson res);
10
11         void onFail(String message);
12     }
13  }
```

因为从后台服务器获取数据属于耗时操作,所以在进行网络请求时需要让 NetIf 在子线程执行,不能让它阻塞主线程,也就是 UI 线程,因此我们写了一个类,用它执行线程切换的任务。与上面相同,把它放在 net 包下,并命名为 ThreadExecutor,详细代码如程序清单 7-18 所示。

程序清单 7-18:hmos\ch07\01\ClockPhone\entry\src\main\java\com\example\clockphone\net\ThreadExecutor.java

```
1   public class ThreadExecutor {
2       //切换任务到主线程执行
3       public static void runUI(Runnable runnable) {
4           EventRunner eventRunner = EventRunner.getMainEventRunner();
                                                                //切换到主线程
5           EventHandler eventHandler = new EventHandler(eventRunner);
6           eventHandler.postSyncTask(runnable);        //执行任务
7       }
8
9       //在子线程执行任务
10      public static void runBG(Runnable runnable) {
11          EventRunner eventRunner = EventRunner.create(true);
                                                                //开启一个新的线程
12          EventHandler eventHandler = new EventHandler(eventRunner);
13          eventHandler.postTask(runnable, 0, EventHandler.Priority.IMMEDIATE);                                //执行任务
14      }
15  }
```

当切换线程的工具类写好后,就可以正式编写访问服务器的工具类了。该类实现 NetIf 接口并重写接口中的相应方法,并且还含有 doGet() 和 buildParams() 两个方法。调用 doGet() 方法时,传入参数向服务器端发送请求,就能得到响应结构;而 buildParams() 方法用于拼接 URL,详细代码如程序清单 7-19 所示。

程序清单 7-19:hmos\ch07\01\ClockPhone\entry\src\main\java\com\example\clockphone\net\Net.java

```
1   public class Net implements NetIf {
```

```java
2      private NetManager netManager;
3      private static final String URL = "http://abcdd.vaiwan.com";
4
5      public Net() {
6          netManager = NetManager.getInstance(null);
7      }
8
9      //拼接URL
10     public static String buildParams(String url, Map<String, Integer> params) {
11         String URLS = URL + url;
12         if (params == null) return null;
13         StringBuilder builder = new StringBuilder(URLS);
14         boolean isFirst = true;
15         for (String key : params.keySet()) {
16             Integer value = params.get(key);
17             if (key != null && value != null) {
18                 if (isFirst) {//第一次拼接时需拼接?号
19                     isFirst = false;
20                     builder.append("?");
21                 } else {
22                     builder.append("&");
23                 }
24                 builder.append(key).append("=").append(value);
25             }
26         }
27         return builder.toString();
28     }
29
30     //向服务器发送GET请求
31     private void doGet(String finalUrl, NetListener listener) {
32         NetHandle netHandle = netManager.getDefaultNet();
33         HttpURLConnection connection = null;
34         InputStream inputStream = null;
35         ByteArrayOutputStream byteArrayOutputStream = null;
36         try {
37             URL url = new URL(finalUrl);
38             URLConnection urlConnection = netHandle.openConnection(url, Proxy.NO_PROXY);
39             if (urlConnection instanceof HttpURLConnection) {
40                 connection = (HttpURLConnection) urlConnection;
41             }
42             connection.setRequestMethod("GET");
```

```
43            connection.connect();
44            if (connection.getResponseCode() == 200) {
45                inputStream = connection.getInputStream();
46                byteArrayOutputStream = new ByteArrayOutputStream();
47                int readLen;
48                byte[] bytes = new byte[1024];
49                while ((readLen = inputStream.read(bytes)) != -1) {
50                    byteArrayOutputStream.write(bytes, 0, readLen);
51                }
52                String result = byteArrayOutputStream.toString();
53                ThreadExecutor.runUI(new Runnable() {
54                    @Override
55                    public void run() {
56                        try {
57                            if (!"".equals(result)) {
58                                HiJson hiJson = new HiJson(new JSONArray(result));
59                                listener.onSuccess(hiJson);
60                            }
61                        } catch (JSONException e) {
62                            e.printStackTrace();
63                            listener.onFail("数据解析错误 code:" + e.toString());
64                        }
65                    }
66                });
67            } else {
68                listener.onFail("请求失败 code:" + connection.getResponseCode());
69            }
70        } catch (Exception e) {
71            e.printStackTrace();
72            listener.onFail("请求失败 msg:" + e.toString());
73        } finally {
74            if (connection != null) {
75                connection.disconnect();
76            }
77            if (inputStream != null) {
78                try {
79                    inputStream.close();
80                } catch (IOException e) {
81                    e.printStackTrace();
82                }
```

```
83              }
84              if (byteArrayOutputStream != null) {
85                  try {
86                      byteArrayOutputStream.close();
87                  } catch (IOException e) {
88                      e.printStackTrace();
89                  }
90              }
91          }
92      }
93
94      @Override
95      public void setClock(Map<String, Integer> params, NetListener
    listener) {                                                    //设置闹钟
96          String finalUrl = buildParams("/setNewClock", params);
97          ThreadExecutor.runBG(new Runnable() {
98              @Override
99              public void run() {
100                 doGet(finalUrl, listener);
101             }
102         });
103     }
104
105     @Override
106     public void getAllClock(NetListener listener) {         //获取所有闹钟
107         String finalUrl = URL + "/searchAllClock";
108         ThreadExecutor.runBG(new Runnable() {
109             @Override
110             public void run() {
111                 doGet(finalUrl, listener);
112             }
113         });
114     }
115
116     @Override
117     public void deleteClock(int[] clockid, NetListener listener) {
                                                                //删除闹钟
118         int count = clockid.length;
119         String finalUrl = URL + "/deleteClock?clockid=";
120         String[] finalUrls = new String[count];
121         for (int i = 0; i < count; i++) {
122             finalUrls[i] = finalUrl + clockid[i];
123         }
```

```
124            for (int j = 0; j < count; j++) {
125                int finalJ = j;
126                ThreadExecutor.runBG(new Runnable() {
127                    @Override
128                    public void run() {
129                        doGet(finalUrls[finalJ], listener);
130                    }
131                });
132            }
133        }
134 }
```

这时,还需要在 config.json 文件的 reqPermissions 字段中声明联网所需要的权限,详细代码如程序清单 7-20 所示。

程序清单 7-20:hmos\ch07\01\ClockPhone\entry\src\main\config.json

```
1   "reqPermissions": [
2       {
3           "name": "ohos.permission.INTERNET"
4       },
5       {
6           "name": "ohos.permission.GET_NETWORK_INFO"
7       },
8       {
9           "name": "ohos.permission.SET_NETWORK_INFO"
10      }
11  ]
```

因为我们使用的是 HTTP,而 HarmonyOS 默认是不接受使用 HTTP 的,鉴于此种情况,我们必须手动设置 HarmonyOS 软件接受使用 HTTP,在 config.json 文件中的 deviceConfig 字段中做出如程序清单 7-21 所示的配置。

程序清单 7-21:hmos\ch07\01\ClockPhone\entry\src\main\config.json

```
1   "deviceConfig": {
2       "default": {
3           "network": {
4               "cleartextTraffic": true
5           }
6       }
```

此时就可以在 MainAbilitySlice 类中调用这些工具类设置主界面的显示效果了,详细代码如程序清单 7-22 所示。

程序清单 7-22：hmos\ch07\01\ClockPhone\entry\src\main\java\com\example\clockphone\slice\MainAbilitySlice.java

```java
public class MainAbilitySlice extends AbilitySlice {
    private ListContainer listClockContainer;
    private ClockListItemProvider clockListItemProvider;
    List<Clock> clockList;
    private Net net = new Net();

    @Override
    public void onStart(Intent intent) {
        super.onStart(intent);
        super.setUIContent(ResourceTable.Layout_ability_main);
        initListContainer();
    }

    private void initListContainer() {
        listClockContainer=(ListContainer) this.findComponentById(ResourceTable.Id_list_clock_view);
        clockListItemProvider = new ClockListItemProvider(this);
        net.getAllClock(new NetIf.NetListener() {
            @Override
            public void onSuccess(HiJson res) {
                bindData(res);
            }

            @Override
            public void onFail(String message) {

            }
        });

    }

    private void bindData(HiJson res) {
        HiJson hiJson;
        int count = res.count();
        this.clockList = new ArrayList<>();
        int i = 0;
        while (i < count) {
            hiJson = res.get(i);
            Clock clock = new Clock();
            clock.setClockid(hiJson.value("clockid"));
```

```
40              clock.setYears(hiJson.value("years"));
41              clock.setMonths(hiJson.value("months"));
42              clock.setDays(hiJson.value("days"));
43              clock.setHours(hiJson.value("hours"));
44              clock.setMinutes(hiJson.value("minutes"));
45              clock.setSends(hiJson.value("sends"));
46              clock.setHappened(hiJson.value("happened"));
47              this.clockList.add(clock);
48              i++;
49          }
50          clockListItemProvider.setDataList(this.clockList);
51          listClockContainer.setItemProvider(clockListItemProvider);
52          listClockContainer.setReboundEffect(true);      //回弹效果
53      }
54
55      @Override
56      public void onActive() {
57          super.onActive();
58      }
59
60      @Override
61      public void onForeground(Intent intent) {
62          super.onForeground(intent);
63      }
64  }
```

运行后显示效果如图 7-35 所示。

图 7-35 闹钟显示模块界面效果

7.3.2 添加闹钟模块

添加闹钟是该软件的重要功能之一，当单击应用主界面右上角的"＋"时就会进入添加闹钟模块界面，其界面预览效果如图7-36。

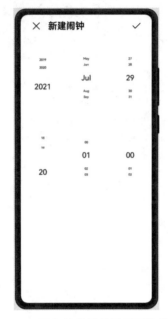

图7-36 添加闹钟模块界面预览效果

添加闹钟模块的布局文件，详细代码如程序清单7-23所示。

程序清单7-23：hmos\ch07\01\ClockPhone\entry\src\main\resources\base\layout\ability_add_clock.xml

```
1   <?xml version="1.0" encoding="utf-8"?>
2   <DirectionalLayout
3       xmlns:ohos="http://schemas.huawei.com/res/ohos"
4       ohos:height="match_parent"
5       ohos:width="match_parent"
6       ohos:orientation="vertical">
7
8       <DependentLayout
9           ohos:height="60vp"
10          ohos:width="match_parent"
11          ohos:padding="10vp">
12
13          <Image
14              ohos:id="$+id:image_cancel"
15              ohos:height="25vp"
```

```
16              ohos:width="25vp"
17              ohos:foreground_element="$media:cancel"
18              ohos:left_margin="20vp"
19              ohos:vertical_center="true"/>
20
21          <Text
22              ohos:id="$+id:text_clockname"
23              ohos:height="match_content"
24              ohos:width="match_content"
25              ohos:background_element="$graphic:background_ability_main"
26              ohos:layout_alignment="horizontal_center"
27              ohos:left_margin="20vp"
28              ohos:right_of="$id:image_cancel"
29              ohos:text="新建闹钟"
30              ohos:text_size="25fp"
31              ohos:vertical_center="true"/>
32
33          <Image
34              ohos:id="$+id:image_ok"
35              ohos:height="25vp"
36              ohos:width="25vp"
37              ohos:align_parent_right="true"
38              ohos:foreground_element="$media:ok"
39              ohos:right_margin="20vp"
40              ohos:vertical_center="true"/>
41      </DependentLayout>
42
43      <DatePicker
44          ohos:id="$+id:date_pick"
45          ohos:height="match_content"
46          ohos:width="match_parent"
47          ohos:operated_text_color="#FF9912"
48          ohos:selected_normal_text_margin_ratio="5"
49          ohos:selected_text_color="#007DFF"
50          ohos:selected_text_size="20fp"
51          ohos:top_padding="50vp"/>
52
53      <TimePicker
54          ohos:id="$+id:time_picker"
55          ohos:height="match_content"
56          ohos:width="match_parent"
57          ohos:operated_text_color="#FF9912"
58          ohos:second="0"
```

```
59          ohos:selected_normal_text_margin_ratio="5"
60          ohos:selected_text_color="#007DFF"
61          ohos:selected_text_size="20fp"
62          ohos:top_padding="90vp"/>
63  </DirectionalLayout>
```

布局文件写好后，接下来在 slice 包下新建一个 AbilitySlice 文件，把它命名为 AddClockAbilitySlice，详细代码如程序清单 7-24 所示。

程序清单 7-24：hmos\ch07\01\ClockPhone\entry\src\main\java\com\example\clockphone\slice\AddClockAbilitySlice.java

```
1   public class AddClockAbilitySlice extends AbilitySlice {
2       private Net net = new Net();
3       private DatePicker datePicker;
4       private TimePicker timePicker;
5       private Integer years, months, days, hours, minutes;
6       private Integer clockid;
7       private Image image_ok, image_cancel;
8
9       @Override
10      protected void onStart(Intent intent) {
11          super.onStart(intent);
12          super.setUIContent(ResourceTable.Layout_ability_add_clock);
13          initLayout();
14          setListener();                                          //设置监听器
15      }
16
17      private void setListener() {
18          image_ok.setClickedListener(new Component.ClickedListener() {
19              @Override
20              public void onClick(Component component) {
21                  setClock();
22              }
23          });
24          image_cancel.setClickedListener(new Component.ClickedListener() {
25              @Override
26              public void onClick(Component component) {
27                  terminate();                                    //返回主界面
28              }
29          });
30      }
31
```

```
32      private void initLayout() {
33          datePicker = (DatePicker) findComponentById(ResourceTable.Id_
    date_pick);
34          timePicker = (TimePicker) findComponentById(ResourceTable.Id_
    time_picker);
35          timePicker.showSecond(false);                    //隐藏秒
36          timePicker.enableSecond(false);                  //设置秒 selector 无法滚动
37          image_ok = (Image) findComponentById(ResourceTable.Id_image_ok);
38          image_cancel = (Image) findComponentById(ResourceTable.Id_image_
    cancel);
39      }
40
41      private Integer getDateAndTimeThentoClockId() { //生成闹钟 ID
42          years = datePicker.getYear();
43          months = datePicker.getMonth();
44          days = datePicker.getDayOfMonth();
45          hours = timePicker.getHour();
46          minutes = timePicker.getMinute();
47          String year = null, month = null, day = null, hour = null, minute = null;
48          year = "" + (years - 2000);
49          if (months < 10) {
50              month = "0" + months;
51          } else {
52              month = "" + minutes;
53          }
54          if (days < 10) {
55              day = "0" + days;
56          } else {
57              day = "" + days;
58          }
59          if (hours < 10) {
60              hour = "0" + hours;
61          } else {
62              hour = "" + hours;
63          }
64          if (minutes < 10) {
65              minute = "0" + minutes;
66          } else {
67              minute = "" + minutes;
68          }
69          String idString = year + month + day + hour + minute;
70          clockid = Integer.parseInt(idString);
71          return clockid;
```

```
72      }
73
74      private void setClock() {
75          Map<String, Integer> params = new HashMap<>();
76          params.put("clockid", getDateAndTimeThentoClockId());
77          params.put("years", years);
78          params.put("months", months);
79          params.put("days", days);
80          params.put("hours", hours);
81          params.put("minutes", minutes);
82          net.setClock(params, new NetIf.NetListener() {
83              @Override
84              public void onSuccess(HiJson res) {
85
86              }
87
88              @Override
89              public void onFail(String message) {
90
91              }
92          });
93          terminate();                                              //返回主界面
94      }
95  }
```

接下来，在 MainAbilitySlice 中加入跳转语句，使得当用户单击软件主界面右上角的"＋"后能跳转到闹钟添加界面，我们在 MainAbilitySlice 中的 onStart() 方法中加入如程序清单 7-25 所示的代码实现跳转操作。

程序清单 7-25：hmos\ch07\01\ClockPhone\entry\src\main\java\com\example\clockphone\slice\MainAbilitySlice.java

```
1   imageAddClock = (Image) this.findComponentById(ResourceTable.Id_image_
    add_clock);
2   imageAddClock.setClickedListener(new Component.ClickedListener() {
3       @Override
4       public void onClick(Component component) {
5           present(new AddClockAbilitySlice(), new Intent());
6       }
7   });
```

当从添加闹钟界面返回主界面后，我们希望主界面的闹钟列表能够更新，所以在 MainAbilitySlice 中编写了一个 refreshClocks() 方法，用于刷新列表数据并在 onForeground() 中调用该方法实现返回刷新操作，详细代码如程序清单 7-26 所示。

程序清单 7-26：hmos\ch07\01\ClockPhone\entry\src\main\java\com\example\clockphone\slice\MainAbilitySlice.java

```
1   private void refreshClocks() {                      //刷新闹钟列表
2       net.getAllClock(new NetIf.NetListener() {
3           @Override
4           public void onSuccess(HiJson res) {
5               bindData(res);
6           }
7
8           @Override
9           public void onFail(String message) {
10
11          }
12      });
13  }
14
15  @Override
16  public void onForeground(Intent intent) {
17      super.onForeground(intent);
18      refreshClocks();                                //返回时刷新闹钟列表
19  }
```

7.3.3 删除闹钟模块

闹钟既要能添加，也要能删除已响过的闹钟信息或不需要的闹钟信息，用户通过长按闹钟列表进入删除界面，在删除界面中，用户可以单独勾选需要删除的闹钟，也可以直接全选，删除所有闹钟。单击"删除"按钮时，会弹出一个消息框，提示用户是否删除。删除闹钟模块界面预览效果如图 7-37 所示。

图 7-37　删除闹钟模块界面预览效果

删除闹钟界面的布局文件详细代码如程序清单 7-27 所示。

程序清单 7-27：ohos\ch07\01\ClockProject工程\ecx=o=c=ccc\csv\base\layout\ability_delete_clock.xml

```xml
1   <?xml version="1.0" encoding="utf-8"?>
2   <DependentLayout
3       xmlns:ohos="http://schemas.huawei.com/res/ohos"
4       ohos:height="match_parent"
5       ohos:width="match_parent"
6       ohos:orientation="vertical">
7
8       <DependentLayout
9           ohos:id="$+id:layout_top"
10          ohos:height="60vp"
11          ohos:width="match_parent"
12          ohos:padding="10vp">
13
14          <Image
15              ohos:id="$+id:image_cancel"
16              ohos:height="25vp"
17              ohos:width="25vp"
18              ohos:foreground_element="$media:cancel"
19              ohos:left_margin="20vp"
20              ohos:vertical_center="true"/>
21
22          <Text
23              ohos:id="$+id:text_clockname"
24              ohos:height="match_content"
25              ohos:width="match_content"
26              ohos:background_element="$graphic:background_ability_main"
27              ohos:layout_alignment="horizontal_center"
28              ohos:left_margin="20vp"
29              ohos:right_of="$id:image_cancel"
30              ohos:text="删除闹钟"
31              ohos:text_size="25fp"
32              ohos:vertical_center="true"/>
33
34      </DependentLayout>
35
36      <ListContainer
37          ohos:id="$+id:list_clock_view"
38          ohos:height="match_content"
39          ohos:width="match_parent"
```

```
40              ohos:below="$id:layout_top"/>
41
42      <DirectionalLayout
43          ohos:height="60vp"
44          ohos:width="match_parent"
45          ohos:align_parent_bottom="true"
46          ohos:layout_alignment="horizontal_center"
47          ohos:orientation="horizontal"
48          ohos:padding="10vp">
49
50          <Button
51              ohos:id="$+id:button_delete"
52              ohos:height="match_content"
53              ohos:width="100vp"
54              ohos:element_left="$media:delete"
55              ohos:left_margin="50vp"
56              ohos:text="删除"
57              ohos:text_size="25fp"/>
58
59          <Checkbox
60              ohos:id="$+id:checkbox_selectall"
61              ohos:height="30vp"
62              ohos:width="100vp"
63              ohos:check_element="$graphic:checkbox_deleteclock"
64              ohos:left_margin="50vp"
65              ohos:text="全选"
66              ohos:text_size="25fp"/>
67      </DirectionalLayout>
68  </DependentLayout>
```

ListContainer 中的每条闹钟信息的布局文件详细代码如程序清单 7-28 所示。

```
1   <?xml version="1.0" encoding="utf-8"?>
2   <DependentLayout
3       xmlns:ohos="http://schemas.huawei.com/res/ohos"
4       ohos:height="match_parent"
5       ohos:width="match_parent"
6       ohos:orientation="vertical">
7
8       <Text
9           ohos:id="$+id:item_text_date"
```

```
10          ohos:height="match_content"
11          ohos:width="match_content"
12          ohos:layout_alignment="center"
13          ohos:left_margin="20vp"
14          ohos:padding="5vp"
15          ohos:text="Date"
16          ohos:text_size="20fp"/>
17
18      <Text
19          ohos:id="$+id:item_text_time"
20          ohos:height="match_content"
21          ohos:width="match_content"
22          ohos:below="$id:item_text_date"
23          ohos:layout_alignment="center"
24          ohos:left_margin="20vp"
25          ohos:padding="5vp"
26          ohos:text="Time"
27          ohos:text_size="20fp"/>
28
29      <Checkbox
30          ohos:id="$+id:item_checkbox"
31          ohos:height="30vp"
32          ohos:width="60vp"
33          ohos:align_parent_right="true"
34          ohos:check_element="$graphic:checkbox_deleteclock"
35          ohos:right_margin="20vp"
36          ohos:top_margin="20vp"/>
37
38      <Component
39          ohos:id="$+id:item_divider"
40          ohos:height="1vp"
41          ohos:width="match_parent"
42          ohos:background_element="$graphic:list_divider"
43          ohos:below="$id:item_text_time"/>
44  </DependentLayout>
```

接下来为 ListContainer 写一个适配器,控制数据的显示。在 provider 包下新建一个 DeleteClockItemProvider 类,该类继承 BaseItemProvider,详细代码如程序清单 7-29 所示。

程序清单 7-29: hmos\ch07\04\ClockPhone\entry\src\main\java\com\example\clockphone\provider\DeleteClockItemProvider.java

```
1   public class DeleteClockItemProvider extends BaseItemProvider {
```

```java
2       private AbilitySlice slice;
3       private List<Clock> dataList = new ArrayList<>();
4       private SimpleDateFormat simpleDateFormat1 = new SimpleDateFormat
    ("HH:mm");
5        private SimpleDateFormat simpleDateFormat2 = new SimpleDateFormat("
    yyyy-MM-dd");
6
7       public DeleteClockItemProvider(AbilitySlice abilitySlice) {
8           this.slice = abilitySlice;
9       }
10
11      @Override
12      public int getCount() {
13          return dataList.size();
14      }
15
16      @Override
17      public Object getItem(int i) {
18          return dataList.get(i);
19      }
20
21      @Override
22      public long getItemId(int i) {
23          return i;
24      }
25
26      public List<Clock> getDataList() {
27          return dataList;
28      }
29
30      public void setDataList(List<Clock> dataList) {
31          this.dataList = dataList;
32      }
33
34      @Override
35      public Component getComponent(int i, Component component,
    ComponentContainer componentContainer) {
36          final Component component1;
37          ViewHolder viewHolder;
38          Clock clock = dataList.get(i);
39          if (component == null) {
40              component1 = LayoutScatter.getInstance(slice).parse
    (ResourceTable.Layout_item_delete_clock, null, false);
```

```
41          viewHolder = new ViewHolder();
42          viewHolder.textDate = (Text) component1.findComponentById
    (ResourceTable.Id_item_text_date);
43          viewHolder.textTime = (Text) component1.findComponentById
    (ResourceTable.Id_item_text_time);
44          viewHolder.checkboxDelete = (Checkbox) component1.
    findComponentById(ResourceTable.Id_item_checkbox);
45          viewHolder.checkboxDelete.setCheckedStateChangedListener
    (new AbsButton.CheckedStateChangedListener() {
46              @Override
47              public void onCheckedChanged(AbsButton absButton, boolean
    isCheck) {
48                  if (isCheck) {
49                      clock.setSends(2);
50                  }
51              }
52          });
53          component1.setTag(viewHolder);
54      } else {
55          component1 = component;
56          viewHolder = (ViewHolder) component.getTag();
57      }
58      Calendar calendar = Calendar.getInstance();
59      calendar.set(Calendar.YEAR, clock.getYears());
60      calendar.set(Calendar.MONTH, clock.getMonths() - 1);
61      calendar.set(Calendar.DAY_OF_MONTH, clock.getDays());
62      calendar.set(Calendar.HOUR_OF_DAY, clock.getHours());
63      calendar.set(Calendar.MINUTE, clock.getMinutes());
64      viewHolder.textTime.setText(simpleDateFormat1.format(calendar.
    getTime()));
65      viewHolder.textDate.setText(simpleDateFormat2.format(calendar.
    getTime()));
66      calendar.clear();
67      if (clock.getSends() == 2) {
68          viewHolder.checkboxDelete.setChecked(true);
69      }
70      return component1;
71  }
72
73  class ViewHolder {
74      Text textDate;
75      Text textTime;
76      Checkbox checkboxDelete;
77  }
78 }
```

接下来，在 slice 包下新建一个名为 DeleteClockAbilitySlice 的 AbilitySlice，用于删除闹钟的逻辑以及页面显示等，详细代码如程序清单 7-30 所示。

程序清单 7-30：hmos\ch07\01\ClockPhone\entry\src\main\java\com\example\clockphone\slice\DeleteClockAbilitySlice.java

```java
 1  public class DeleteClockAbilitySlice extends AbilitySlice {
 2      private ListContainer listContainer;
 3      private Checkbox checkboxSelectAll;
 4      private DeleteClockItemProvider listViewItemProvider;
 5      private Net net = new Net();
 6  
 7      @Override
 8      protected void onStart(Intent intent) {
 9          super.onStart(intent);
10          super.setUIContent(ResourceTable.Layout_ability_delete_clock);
11          initListContainer();
12          initView();
13      }
14  
15      private void initListContainer() {
16          listContainer=(ListContainer) this.findComponentById(ResourceTable.Id_list_clock_view);
17          listViewItemProvider = new DeleteClockItemProvider(this);
18          net.getAllClock(new NetIf.NetListener() {
19              @Override
20              public void onSuccess(HiJson res) {
21                  bindData(res);
22              }
23  
24              @Override
25              public void onFail(String message) {
26  
27              }
28          });
29      }
30  
31      private void initView() {
32          Image imageCancel=(Image) findComponentById(ResourceTable.Id_image_cancel);
33          imageCancel.setClickedListener(new Component.ClickedListener() {
34              @Override
35              public void onClick(Component component) {
36                  terminate();
```

```
37          }
38      });
39
40      checkboxSelectAll=(Checkbox) findComponentById(ResourceTable.
    Id_checkbox_selectall);
41      checkboxSelectAll.setCheckedStateChangedListener((absButton,
    isCheck) -> {
42          List<Clock> clockList = listViewItemProvider.getDataList();
43          for (Clock clock : clockList) {
44              clock.setSends(2);
45          }
46          listViewItemProvider.notifyDataChanged();
47      });
48
49      Button btnDelete = (Button) findComponentById(ResourceTable.Id_
    button_delete);
50      btnDelete.setClickedListener(new Component.ClickedListener() {
51          @Override
52          public void onClick(Component component) {
53              List<Clock> clockList = listViewItemProvider.getDataList();
54              int selectClock = 0;
55              for (Clock clock : clockList) {
56                  if (clock.getSends() == 2) {
57                      selectClock++;
58                  }
59              }
60              if (selectClock > 0) {
61                  CommonDialog alertDialog = new CommonDialog
    (DeleteClockAbilitySlice.this);
62                  TextField textFieldName = new TextField
    (DeleteClockAbilitySlice.this);
63                  textFieldName.setText(String.format("是否删%d个闹
    钟?", selectClock));
64                  textFieldName.setTextSize(20, Text.TextSizeType.FP);
65                  DirectionalLayout.LayoutConfig layoutConfig = new
    DirectionalLayout.LayoutConfig();
66                  layoutConfig.alignment = LayoutAlignment.CENTER;
67                  textFieldName.setLayoutConfig(layoutConfig);
68                  alertDialog.setContentCustomComponent
    (textFieldName);
69                  alertDialog.setButton(IDialog.BUTTON1, "取消", new
    IDialog.ClickedListener() {
70                      @Override
```

```
71                     public void onClick(IDialog iDialog, int i) {
72                         iDialog.destroy();
73                     }
74                 });
75                 alertDialog.setButton(IDialog.BUTTON3, "删除", new
    IDialog.ClickedListener() {
76                     @Override
77                     public void onClick(IDialog iDialog, int i) {
78                         List<Clock> clockDeleteList =
    listViewItemProvider.getDataList();
79                         int count = 0;
80                         for (Clock clock : clockDeleteList) {
81                             if (clock.getSends() == 2) {
82                                 count++;
83                             }
84                         }
85                         int[] clockDeleteInts = new int[count];
86                         List<Clock> clockDeleteLists = new ArrayList<>();
87                         int j = 0;
88                         for (Clock clock : clockDeleteList) {
89                             if (clock.getSends() == 2) {
90                                 clockDeleteLists.add(clock);
91                                 clockDeleteInts[j] = clock.getClockid();
92                                 j++;
93                             }
94                         }
95                         net.deleteClock(clockDeleteInts, new NetIf.
    NetListener() {
96                             @Override
97                             public void onSuccess(HiJson res) {
98
99                             }
100
101                            @Override
102                            public void onFail(String message) {
103
104                            }
105                        });
106                        iDialog.destroy();
107                        net.getAllClock(new NetIf.NetListener() {
108                            @Override
109                            public void onSuccess(HiJson res) {
110                                bindData(res);
```

```
111                             }
112
113                             @Override
114                             public void onFail(String message) {
115
116                             }
117                         });
118                     }
119                 });
120                 alertDialog.show();
121             }
122         }
123     });
124 }
125
126 private void bindData(HiJson res) {
127     HiJson hiJson;
128     int count = res.count();
129     List<Clock> clockLists = new ArrayList<>();
130     int i = 0;
131     while (i < count) {
132         hiJson = res.get(i);
133         Clock clock = new Clock();
134         clock.setClockid(hiJson.value("clockid"));
135         clock.setYears(hiJson.value("years"));
136         clock.setMonths(hiJson.value("months"));
137         clock.setDays(hiJson.value("days"));
138         clock.setHours(hiJson.value("hours"));
139         clock.setMinutes(hiJson.value("minutes"));
140         clock.setSends(hiJson.value("sends"));
141         clock.setHappened(hiJson.value("happened"));
142         clockLists.add(clock);
143         i++;
144     }
145     listViewItemProvider.setDataList(clockLists);
146     listContainer.setItemProvider(listViewItemProvider);
147     listViewItemProvider.notifyDataChanged();
148     listContainer.setItemProvider(listViewItemProvider);
149     checkboxSelectAll.setChecked(false);
150 }
151 }
```

现在需要在 MainAbilitySlice 中设置当用户长按闹钟列表后跳转到删除闹钟界面的操作。在 initListContainer()方法中加入如程序清单 7-31 所示代码，实现长按跳转的效果。

程序清单 7-31 hmos\ch07\0f\ClockPhone\entry\src\main\java\com\example\clockphone\slice\MainAbilitySlice.java

```
1   listClockContainer.setItemLongClickedListener(new ListContainer.
    ItemLongClickedListener() {
2       @Override
3       public boolean onItemLongClicked(ListContainer listContainer,
    Component component, int i, long l) {
4           present(new DeleteClockAbilitySlice(), new Intent());
5           return true;
6       }
7   });
```

删除闹钟模块的代码均编写完成后，我们就来运行项目看一下效果，如图 7-38(a)所示，当长按闹钟列表时，会跳转到删除闹钟界面。勾选第一个闹钟，然后单击左下角的"删除"按钮进行删除，此时会弹出对话框，提示是否需要删除该闹钟[图 7-38(b)]，单击"删除"按钮后就会进行删除操作。成功删除选中的闹钟后，删除闹钟界面也会进行一次刷新[图 7-38(c)]。如果想返回应用的主界面，可以单击应用左上角的"×"图标返回。

(a) 界面显示效果　　　　(b) 指示显示效果　　　　(c) 删除闹钟成功

图 7-38　删除闹钟模块

7.4 "远程闹钟"手表端应用设计

7.4.1 手表端应用的创建

手机端应用设计完后,紧接着就是手表端的应用设计。手机端包含一个闹铃功能,我们通过按时拉起内置的音乐来达到闹铃的功能。

新建一个项目,并命名为 ClockWearable。需要注意的是,我们需要选择该项目为 Wearable 项目,如图 7-39 所示。

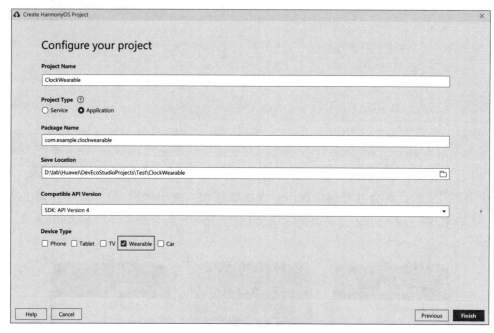

图 7-39 创建 ClockWearable 项目

因为在手表端项目中一样需要联网以及使用到 Clock 的 JavaBean,所以可直接复制手机端项目 ClockPhone 中的 net 包及 bean 包到手表端项目 ClockWearable 中,如图 7-40 所示。由于 Net 和 NetIf 中使用了 HiJson,所以需要导入 HiJson 的 JAR 包,同时还需要在配置文件(config.json)中加入联网的权限以及设置 HTTP 许可,详细代码见 7.3.1 节中的代码。

在手表端应用中不需要设置闹钟以及删除闹钟的功能,我们可以删除 Net.java 和 NetIf.java 中的 setClock() 和 deleteClock() 方法。同时,手表端需要用到设置修改 sends 值和 happened 值的功能,所以我们在 NetIf.java 中添加 setHappened() 和 setSends() 方法并在 Net.java 中重写这两个方法。

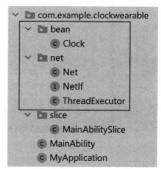

图 7-40 复制代码文件

在 NetIf.java 中添加 setHappened()和 setSends()方法后，NetIf.java 的详细代码如程序清单 7-32 所示。

程序清单 7-32：hmos\ch07\01\ClockWearable\entry\src\main\java\com\example\clockwearable\net\NetIf.java

```
1   public interface NetIf {
2       void getAllClock(NetListener listener);
3
4       void setHappened(Map<String, Integer> params, NetListener listener);
5
6       void setSends(Map<String, Integer> params, NetListener listener);
7
8       interface NetListener {
9           void onSuccess(HiJson res);
10
11          void onFail(String message);
12      }
13  }
```

在 Net.java 中重写 setHappened()和 setSends()方法的详细代码如程序清单 7-33 所示。

程序清单 7-33：hmos\ch07\01\ClockWearable\entry\src\main\java\com\example\clockwearable\net\Net.java

```
1   @Override
2   public void setHappened(Map<String, Integer> params, NetListener listener) {
3       String finalUrl = buildParams("/setHappened", params);
4       ThreadExecutor.runBG(new Runnable() {
5           @Override
6           public void run() {
7               doGet(finalUrl, listener);
8           }
9       });
10  }
11
12  @Override
13  public void setSends(Map<String, Integer> params, NetListener listener) {
14      String finalUrl = buildParams("/setSends", params);
15      ThreadExecutor.runBG(new Runnable() {
16          @Override
17          public void run() {
18              doGet(finalUrl, listener);
```

```
19          }
20      });
21  }
```

接下来,编写手表端的主界面,详细代码如程序清单 7-34 所示。

程序清单 7-34: bmos\ch07\01\ClockWearable\entry\src\main\resources\base\layout\ability_main.xml

```
1   <?xml version="1.0" encoding="utf-8"?>
2   <DirectionalLayout
3       xmlns:ohos="http://schemas.huawei.com/res/ohos"
4       ohos:height="match_parent"
5       ohos:width="match_parent"
6       ohos:alignment="center"
7       ohos:orientation="vertical"
8       ohos:background_element="$media:bg">
9
10      <Image
11          ohos:height="100vp"
12          ohos:width="100vp"
13          ohos:background_element="$media:remind_no"
14          ohos:scale_x="0.5"
15          ohos:scale_y="0.5"/>
16
17      <Text
18          ohos:height="match_content"
19          ohos:width="match_content"
20          ohos:text="还没到时间哦!"
21          ohos:text_color="#00ff77"
22          ohos:text_size="20vp"/>
23
24  </DirectionalLayout>
```

闹钟首页界面预览效果如图 7-41 所示。

图 7-41 闹钟首页界面预览效果

当闹钟需要响起时,应用切换到闹钟闹铃界面,该界面中包含一个 Text 组件以及一个 Button 组件,Button 组件用于提前关闭闹铃,详细代码如程序清单 7-35 所示。

程序清单 7-35：HmOS_c07_01\ClockWearable\entry\src\main\resources\base\layout\ability_clock.xml

```xml
1  <?xml version="1.0" encoding="utf-8"?>
2  <DirectionalLayout
3      xmlns:ohos="http://schemas.huawei.com/res/ohos"
4      ohos:height="match_parent"
5      ohos:width="match_parent"
6      ohos:alignment="center"
7      ohos:orientation="vertical"
8      ohos:background_element="$media:bg_remind">
9
10     <Image
11         ohos:height="100vp"
12         ohos:width="100vp"
13         ohos:image_src="$media:remind"
14         ohos:scale_x="0.5"
15         ohos:scale_y="0.5"/>
16
17     <Text
18         ohos:height="match_content"
19         ohos:width="match_content"
20         ohos:text="该起床了哦！"
21         ohos:text_color="#ff0077"
22         ohos:text_size="20vp"/>
23
24     <Button
25         ohos:id="$+id:btn_close"
26         ohos:height="match_content"
27         ohos:width="match_content"
28         ohos:padding="6vp"
29         ohos:text="关闭闹铃"
30         ohos:text_size="18vp"
31         ohos:top_margin="10fp"
32         ohos:background_element="$graphic:background_button"/>
33 </DirectionalLayout>
```

闹铃界面预览效果如图 7-42 所示。

7.4.2 闹铃定时播放模块

接下来实现闹钟响起的功能。因为是通过播放内置好的音频达到闹铃的效果,所以先编写用于实现音频播放的工具类。新建一个 util 包用来存放这些工具类(图 7-43),其中一个 PlayerManager 实现音频播放管理,另一个 PlayerStateListener 为接口,用于简化构造 IPlayerCallback 后使用不到的抽象方法。IPlayerCallback 是在播放完成、播放位置更改和视频大小更改时提供媒体播放器回调。

图 7-42　闹铃界面预览效果

图 7-43　音频播放相关代码文件

PlayerStateListener.java 的详细代码如程序清单 7-36 所示。

程序清单 7-36:hmos\ch07\01\ClockWearable\entry\src\main\java\com\example\clockwearable\util\PlayerStateListener.java

```
1    public interface PlayerStateListener {
2        void onMusicFinished();
3    }
```

在 PlayerManager 中通过使用视频播放类 Player 实现播放音频效果。Player 类定义的常用操作方法如表 7-2 所示。

表 7-2　Player 类定义的常用操作方法

方法	描述
Player(Context context)	创建 Player 实例
boolean setSource(Source source)	设置媒体源
boolean prepare()	准备播放
boolean play()	开始播放
boolean stop()	停止播放
boolean release()	释放播放资源
boolean isNowPlaying()	检查是否正在播放
boolean rewindTo(long microseconds)	更改播放位置

在 PlayerManager 中我们还构造了 IPlayerCallback,IPlayerCallback 在播放完成、播

放位置更改和视频大小更改时提供媒体播放器回调。IPlayerCallback 接口包含的方法如表 7-3 所示。

表 7-3　IPlayerCallback 接口包含的方法

方　　法	描　　述
void onPrepared()	当媒体文件准备好播放时回调
void onMessage(int type, int extra)	当收到播放器消息或警告时回调
void onError(int errorType, int errorCode)	当收到播放器错误消息时调用
void onResolutionChanged(int width, int height)	当视频尺寸大小更改时调用
void onPlayBackComplete()	播放完成时调用
void onRewindToComplete()	当播放位置通过 Player.rewindTo(long microseconds)更改时调用
void onBufferingChange(int percent)	当缓冲百分比更新时调用
void onMediaTimeIncontinuity(Player.MediaTimeInfo info)	当媒体时间连续性中断时调用，例如：当播放过程中发生错误时，播放位置通过 Player.rewindTo(long microseconds)更改时，或播放速度突然改变时

使用 Player 播放音频时，先创建一个 Player 实例，然后调用 setSource(Source source)方法设置媒体源，之后调用 prepare()方法准备播放，最后调用 play()方法播放设置好的媒体文件。在音频正在播放时，可以调用 stop()方法停止播放，播放结束后，调用 release()方法释放资源。

实现音频播放管理的 PlayerManager 的详细代码如程序清单 7-37 所示。

程序清单 7-37：hmos\ch07\01\ClockWearable\entry\src\main\java\com\example\clockwearable\util\PlayerManager.java

```
1   public class PlayerManager {
2       private Context context;
3       private String playerUri;
4       private Player player;
5       private PlayerStateListener playerStateListener;
6       private boolean isPrepared;
7
8       public PlayerManager(Context context, String playerUri) {
9           this.context = context;
10          this.playerUri = playerUri;
11      }
12
13      public void setPlayerStateListener(PlayerStateListener
    playerStateListener) {
14          this.playerStateListener = playerStateListener;
```

```
15        }
16
17        //实例化
18        public void init() {
19            player = new Player(context);
20            player.setPlayerCallback(new PlayCallBack());
21            setResource(playerUri);
22        }
23
24        //设置资源路径
25        public void setResource(String uri) {
26            try {
27                RawFileEntry rawFileEntry=context.getResourceManager().getRawFileEntry(uri);
28                BaseFileDescriptor baseFileDescriptor=rawFileEntry.openRawFileDescriptor();
29                if (!player.setSource(baseFileDescriptor)) {
30                    return;
31                }
32                isPrepared = player.prepare();
33            } catch (IOException e) {
34                e.printStackTrace();
35            }
36        }
37
38        //播放音频
39        public void play() {
40            if (!isPrepared) {
41                return;
42            }
43            if (!player.play()) {
44                return;
45            }
46            player.play();
47        }
48
49        //停止播放,释放资源
50        public void releasePlayer() {
51            if (player == null) {
52                return;
53            }
54            player.stop();
55            player.release();
```

```
56        }
57
58        //判断是否正在播放
59        public boolean isPlaying() {
60            return player.isNowPlaying();
61        }
62
63        private class PlayCallBack implements Player.IPlayerCallback {
64
65            @Override
66            public void onPrepared() {
67
68            }
69
70            @Override
71            public void onMessage(int i, int i1) {
72
73            }
74
75            @Override
76            public void onError(int i, int i1) {
77
78            }
79
80            @Override
81            public void onResolutionChanged(int i, int i1) {
82
83            }
84
85            @Override
86            public void onPlayBackComplete() {
87                playerStateListener.onMusicFinished();
88            }
89
90            @Override
91            public void onRewindToComplete() {
92
93            }
94
95            @Override
96            public void onBufferingChange(int i) {
97
98            }
```

```
99
100        @Override
101        public void onNewTimedMetaData(Player.MediaTimedMetaData
    mediaTimedMetaData) {
102
103        }
104
105        @Override
106        public void onMediaTimeIncontinuity(Player.MediaTimeInfo
    mediaTimeInfo) {
107
108        }
109    }
110 }
```

写好音频播放工具类后，接下来就是在 AbilitySlice 中使用它们了。在 slice 包下新建一个 AbilitySlice，并命名为 ClockAbilitySlice，该 AbilitySlice 使用 ability_clock.xml 布局文件。ClockAbilitySlice 中包含两个自定义方法：setHappened()方法用于当闹钟响完时将数据库中该条闹钟信息的 happened 值设置为 1，表示该闹钟已经响过；initPlayer()方法用于初始化并使用音频播放工具类。ClockAbilitySlice 的详细代码如程序清单 7-38 所示。

程序清单 7-38：hmos\ch07\01\ClockWearable\entry\src\main\java\com\example\clockwearable\slice\ClockAbilitySlice.java

```
1   public class ClockAbilitySlice extends AbilitySlice {
2       private PlayerManager playerManager;
3       private Net net = new Net();
4       private int clockid;
5
6       @Override
7       protected void onStart(Intent intent) {
8           super.onStart(intent);
9           super.setUIContent(ResourceTable.Layout_ability_clock);
10          clockid = intent.getIntParam("clockid", 0);
11          initPlayer();
12          Button btnClose = (Button) findComponentById(ResourceTable.Id_
    btn_close);
13          btnClose.setClickedListener((button) -> {
14              if (playerManager.isPlaying()) {
15                  playerManager.releasePlayer();
16              }
17              setHappened();
18              this.terminate();                    //暂时直接关闭
```

```java
19            });
20        }
21
22        //当闹钟响完时设置该闹钟的 happened 值为 1
23        private void setHappened() {
24            Map<String, Integer> params = new HashMap<>();
25            params.put("clockid", clockid);
26            try {
27                net.setHappened(params, new NetIf.NetListener() {
28
29                    @Override
30                    public void onSuccess(HiJson res) {
31                    }
32
33                    @Override
34                    public void onFail(String message) {
35                    }
36                });
37            } catch (Exception e) {
38                e.printStackTrace();
39            }
40        }
41
42        private void initPlayer() {
43            playerManager = new PlayerManager(this, "resources/rawfile/Alarm.mp3");
44            playerManager.setPlayerStateListener(new PlayerStateListener() {
45                @Override
46                public void onMusicFinished() {
47                    setHappened();
48                    terminate();
49                }
50            });
51            playerManager.init();
52            playerManager.play();
53        }
54    }
```

上述代码使用 Alarm.mp3 作为闹钟铃声，我们需要在"entry＞src＞main＞resources"目录下新建一个 rawfile 文件夹，在该文件夹下放入一个 Alarm.mp3 文件为闹铃铃声，如图 7-44 所示。

接下来，为了减少联网这个耗时操作的次数，我们通过关系数据库把后台服务器上的闹钟信息存储到本地。基于 Data Ability 对关系数据库进行操作，新建一个 Data

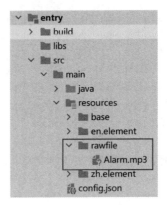

图 7-44　闹铃音频文件

Ability,并命名为ClockDataAbility,详细代码如程序清单7-39所示。

程序清单7-39　hmos-ch07-01-ClockWearable-entry-src-main-java-com-example-clockwearable-ClockDataAbility.java

```
1   public class ClockDataAbility extends Ability {
2       private static final HiLogLabel LABEL_LOG = new HiLogLabel(3,
    0xD001100, "Demo");
3       public static final String DB_NAME = "clockdataability.db";
4       public static final String DB_TAB_NAME = "myclock";
5       public static final String DB_COLUMN_CLOCK_ID = "clockid";
6       public static final String DB_COLUMN_YEAR = "years";
7       public static final String DB_COLUMN_MONTH = "months";
8       public static final String DB_COLUMN_DAY = "days";
9       public static final String DB_COLUMN_HOUR = "hours";
10      public static final String DB_COLUMN_MINUTE = "minutes";
11      public static final String DB_COLUMN_HAPPENED = "happened";
12      private static final int DB_VERSION = 1;
13      private StoreConfig config = StoreConfig.newDefaultConfig(DB_NAME);
14      private RdbStore rdbStore;
15      private RdbOpenCallback rdbOpenCallback = new RdbOpenCallback() {
16          @Override
17          public void onCreate(RdbStore rdbStore) {
18              rdbStore.executeSql("create table if not exists "
19                      + DB_TAB_NAME + " ("
20                      + DB_COLUMN_CLOCK_ID + " integer primary key, "
21                      + DB_COLUMN_YEAR + " integer not null, "
22                      + DB_COLUMN_MONTH + " integer not null, "
23                      + DB_COLUMN_DAY + " integer not null, "
24                      + DB_COLUMN_HOUR + " integer not null, "
25                      + DB_COLUMN_MINUTE + " integer not null, "
```

```
26                    + DB_COLUMN_HAPPENED + " integer not null)");
27          }
28
29          @Override
30          public void onUpgrade(RdbStore rdbStore, int oldVersion, int
    newVersion) {
31
32          }
33      };
34
35      @Override
36      public void onStart(Intent intent) {
37          super.onStart(intent);
38          HiLog.info(LABEL_LOG, "ClockDataAbility onStart");
39          DatabaseHelper databaseHelper = new DatabaseHelper(this);
40          rdbStore = databaseHelper.getRdbStore(config, DB_VERSION,
    rdbOpenCallback, null);
41      }
42
43      @Override
44      public ResultSet query(Uri uri, String[] columns,
    DataAbilityPredicates predicates) {
45          RdbPredicates rdbPredicates=DataAbilityUtils.
    createRdbPredicates(predicates, DB_TAB_NAME);
46          ResultSet resultSet = rdbStore.query(rdbPredicates, columns);
47          if (resultSet == null) {
48              HiLog.info(LABEL_LOG, "resultSet is null");
49          }
50          return resultSet;
51      }
52
53      @Override
54      public int insert(Uri uri, ValuesBucket value) {
55          HiLog.info(LABEL_LOG, "ClockDataAbility insert");
56          String path = uri.getLastPath();
57          if (!DB_TAB_NAME.equals(path)) {
58              HiLog.info(LABEL_LOG, "DataAbility insert path is not matched");
59              return -1;
60          }
61          ValuesBucket values = new ValuesBucket();
62          values.putInteger(DB_COLUMN_CLOCK_ID, value.getInteger(DB_
    COLUMN_CLOCK_ID));
63          values.putInteger(DB_COLUMN_YEAR, value.getInteger(DB_COLUMN_
    YEAR));
```

```
64          values.putInteger(DB_COLUMN_MONTH, value.getInteger(DB_COLUMN_
    MONTH));
65          values.putInteger(DB_COLUMN_DAY, value.getInteger(DB_COLUMN_
    DAY));
66          values.putInteger(DB_COLUMN_HOUR, value.getInteger(DB_COLUMN_
    HOUR));
67          values.putInteger(DB_COLUMN_MINUTE, value.getInteger(DB_COLUMN_
    MINUTE));
68          values.putInteger(DB_COLUMN_HAPPENED, value.getInteger(DB_
    COLUMN_HAPPENED));
69          int index = (int) rdbStore.insert(DB_TAB_NAME, values);
70          DataAbilityHelper.creator(this, uri).notifyChange(uri);
71          return index;
72      }
73
74      @Override
75      public int delete(Uri uri, DataAbilityPredicates predicates) {
76          RdbPredicates rdbPredicates = DataAbilityUtils.
    createRdbPredicates(predicates, DB_TAB_NAME);
77          int index = rdbStore.delete(rdbPredicates);
78          DataAbilityHelper.creator(this, uri).notifyChange(uri);
79          return index;
80      }
81
82      @Override
83      public int update(Uri uri, ValuesBucket value, DataAbilityPredicates
    predicates) {
84          RdbPredicates rdbPredicates = DataAbilityUtils.
    createRdbPredicates(predicates, DB_TAB_NAME);
85          int index = rdbStore.update(value, rdbPredicates);
86          DataAbilityHelper.creator(this, uri).notifyChange(uri);
87          return index;
88      }
89
90      @Override
91      public FileDescriptor openFile(Uri uri, String mode) {
92          return null;
93      }
94
95      @Override
96      public String[] getFileTypes(Uri uri, String mimeTypeFilter) {
97          return new String[0];
98      }
```

```
99
100     @Override
101     public PacMap call(String method, String arg, PacMap extras) {
102         return null;
103     }
104
105     @Override
106     public String getType(Uri uri) {
107         return null;
108     }
109 }
```

那么,怎么间断从服务器端取到最新的数据存储到本地并且按时拉起闹钟呢?可以通过 Service Ability 实现这一功能。新建一个 Service Ability,并命名为 ClockServiceAbility,详细代码如程序清单 7-40 所示。

程序清单 7-40 hmos\ch07\01\ClockWearable\entry\src\main\java\com\example\clockwearable\ClockServiceAbility.java

```
1   public class ClockServiceAbility extends Ability {
2       private static final HiLogLabel LABEL_LOG=new HiLogLabel(3,
    0xD001100, "Demo");
3       private Net net = new Net();
4       private static final String DB_COLUMN_CLOCK_ID = "clockid";
5       private static final String DB_COLUMN_YEAR = "years";
6       private static final String DB_COLUMN_MONTH = "months";
7       private static final String DB_COLUMN_DAY = "days";
8       private static final String DB_COLUMN_HOUR = "hours";
9       private static final String DB_COLUMN_MINUTE = "minutes";
10      private static final String DB_COLUMN_SEND = "sends";
11      private static final String DB_COLUMN_HAPPENED = "happened";
12      private DataAbilityHelper databaseHelper;
13      private static final String BASE_URI="dataability:///com.example.
    clockwearable.ClockDataAbility";
14      private static final String DATA_PATH = "/myclock";
15      private boolean isInAlarm = false;
16      private int clockid;
17
18      @Override
19      public void onStart(Intent intent) {
20          HiLog.error(LABEL_LOG, "ClockServiceAbility::onStart");
21          super.onStart(intent);
22          databaseHelper = DataAbilityHelper.creator(this);
```

```
23          ScheduledExecutorService scheduledExecutorService = Executors.
   newScheduledThreadPool(1);
24          scheduledExecutorService.schedule(new Runnable() {
25              @Override
26              public void run() {
27                  getClock();                    //从后台获取闹钟
28                  doClock();                     //按时拉起闹钟
29              }
30          }, 10, TimeUnit.SECONDS);
31      }
32
33      //判断是否有闹钟需要拉起
34      private void doClock() {
35          if (isInAlarm) {
36              return;
37          }
38          Calendar calendar = Calendar.getInstance();
39          int day = calendar.get(Calendar.DAY_OF_MONTH);
40          int hour = calendar.get(Calendar.HOUR_OF_DAY);
41          int minute = calendar.get(Calendar.MINUTE);
42          List<Clock> clockList = getLocalAllClocks();
43          for (Clock clock : clockList) {
44              if (clock.getHappened() == 0 && clock.getDays() == day &&
   clock.getHours() == hour && clock.getMinutes() == minute) {
45                  clockid = clock.getClockid();
46                  startClockAlarmAbility(clockid);
47              }
48          }
49      }
50
51      //启动闹钟
52      private void startClockAlarmAbility(int clockid) {
53          isInAlarm = true;
54          Intent intent = new Intent();
55          Operation operation = new Intent.OperationBuilder()
56                  .withBundleName(getBundleName())
57                  .withAbilityName(ClockAbilitySlice.class.getName())
58                  .build();
59          intent.setOperation(operation);
60          intent.setParam("clockid", clockid);
61          startAbility(intent);
62      }
63
```

```
64          //获取本地闹钟数据
65          private List<Clock> getLocalAllClocks() {
66              List<Clock> clockList = new LinkedList<>();
67              String[] columns = new String[]{ClockDataAbility.DB_COLUMN_CLOCK_
    ID, ClockDataAbility.DB_COLUMN_YEAR,
68                      ClockDataAbility.DB_COLUMN_MONTH, ClockDataAbility.DB_
    COLUMN_DAY, ClockDataAbility.DB_COLUMN_HOUR,
69                      ClockDataAbility.DB_COLUMN_MINUTE, ClockDataAbility.DB_
    COLUMN_HAPPENED};
70              DataAbilityPredicates predicates = new DataAbilityPredicates();
71              try {
72                  ResultSet resultSet=databaseHelper.query(Uri.parse(BASE_URI +
    DATA_PATH), columns, predicates);
73                  if (resultSet == null || resultSet.getRowCount() == 0) {
74                      return clockList;
75                  }
76                  resultSet.goToFirstRow();
77                  do {
78                      Clock clock = new Clock();
79                      clock.setClockid(resultSet.getInt(resultSet.
    getColumnIndexForName(ClockDataAbility.DB_COLUMN_CLOCK_ID)));
80                      clock.setYears(resultSet.getInt(resultSet.
    getColumnIndexForName(ClockDataAbility.DB_COLUMN_YEAR)));
81                      clock.setMonths(resultSet.getInt(resultSet.
    getColumnIndexForName(ClockDataAbility.DB_COLUMN_MONTH)));
82                      clock.setDays(resultSet.getInt(resultSet.
    getColumnIndexForName(ClockDataAbility.DB_COLUMN_DAY)));
83                      clock.setHours(resultSet.getInt(resultSet.
    getColumnIndexForName(ClockDataAbility.DB_COLUMN_HOUR)));
84                      clock.setMinutes(resultSet.getInt(resultSet.
    getColumnIndexForName(ClockDataAbility.DB_COLUMN_MINUTE)));
85                      clock.setHappened(resultSet.getInt(resultSet.
    getColumnIndexForName(ClockDataAbility.DB_COLUMN_HAPPENED)));
86                      clockList.add(clock);
87                  } while (resultSet.goToNextRow());
88              } catch (DataAbilityRemoteException | IllegalStateException
    exception) {
89                  exception.printStackTrace();
90              }
91              return clockList;
92          }
93
94          //从后台服务器获取闹钟数据
```

```java
 95        private void getClock() {
 96            net.getAllClock(new NetIf.NetListener() {
 97                @Override
 98                public void onSuccess(HiJson res) {
 99                    checkClockAndInsert(res);//保存未保存到本地的闹钟到本地
100                }
101
102                @Override
103                public void onFail(String message) {
104
105                }
106            });
107        }
108
109        //检查查询到的数据,判断其是否已被保存到本地或已响过
110        private void checkClockAndInsert(HiJson res) {
111            HiJson hiJson;
112            int count = res.count();
113            int i = 0;
114            int happened, sends;
115            while (i < count) {
116                hiJson = res.get(i);
117                happened = hiJson.value(DB_COLUMN_HAPPENED);
118                sends = hiJson.value(DB_COLUMN_SEND);
119                if (happened == 0 && sends == 0) {
120                    inserts(hiJson.value(DB_COLUMN_CLOCK_ID), hiJson.value
    (DB_COLUMN_YEAR), hiJson.value(DB_COLUMN_MONTH), hiJson.value(DB_COLUMN
    _DAY), hiJson.value(DB_COLUMN_HOUR), hiJson.value(DB_COLUMN_MINUTE),
    hiJson.value(DB_COLUMN_HAPPENED));
121                    Map<String, Integer> params = new HashMap<>();
122                    params.put("clockid", hiJson.value(DB_COLUMN_CLOCK_ID));
123                    net.setSends(params, new NetIf.NetListener() {
                                         //修改已保存的闹钟的 sends 值为 1
124                        @Override
125                        public void onSuccess(HiJson res) {
126
127                        }
128
129                        @Override
130                        public void onFail(String message) {
131
132                        }
133                    });
```

```
134            }
135            i++;
136        }
137    }
138
139    //插入数据到本地
140    private void inserts(int clockid, int years, int months, int days, int
   hours, int minutes, int happened) {
141        ValuesBucket valuesBucket = new ValuesBucket();
142        valuesBucket.putInteger(DB_COLUMN_CLOCK_ID, clockid);
143        valuesBucket.putInteger(DB_COLUMN_YEAR, years);
144        valuesBucket.putInteger(DB_COLUMN_MONTH, months);
145        valuesBucket.putInteger(DB_COLUMN_DAY, days);
146        valuesBucket.putInteger(DB_COLUMN_HOUR, hours);
147        valuesBucket.putInteger(DB_COLUMN_MINUTE, minutes);
148        valuesBucket.putInteger(DB_COLUMN_HAPPENED, happened);
149        try {
150            if (databaseHelper.insert(Uri.parse(BASE_URI + DATA_PATH),
   valuesBucket) != -1) {
151                HiLog.debug(LABEL_LOG, "insert successful");
152            } else {
153                HiLog.debug(LABEL_LOG, "insert successful?????");
154            }
155        } catch (DataAbilityRemoteException | IllegalStateException
   exception) {
156            exception.printStackTrace();
157            HiLog.debug(LABEL_LOG, "insert: dataRemote exception|
   illegalStateException");
158            HiLog.debug(LABEL_LOG, exception.toString());
159        }
160    }
161
162    @Override
163    public void onBackground() {
164        super.onBackground();
165        HiLog.info(LABEL_LOG, "ClockServiceAbility::onBackground");
166    }
167
168    @Override
169    public void onStop() {
170        super.onStop();
171        HiLog.info(LABEL_LOG, "ClockServiceAbility::onStop");
172    }
```

```
173
174      @Override
175      public void onCommand(Intent intent, boolean restart, int startId) {
176      }
177
178      @Override
179      public IRemoteObject onConnect(Intent intent) {
180          return null;
181      }
182
183      @Override
184      public void onDisconnect(Intent intent) {
185      }
186  }
```

接下来,只要在 MainAbilitySlice 启动 ClockServiceAbility 即可,详细代码如程序清单 7-41 所示。

程序清单 7-41: hmos\ch07\01\ClockWearable\entry\src\main\java\com\example\clockwearable\slice\MainAbilitySlice.java

```
1   public class MainAbilitySlice extends AbilitySlice {
2
3       @Override
4       public void onStart(Intent intent) {
5           super.onStart(intent);
6           super.setUIContent(ResourceTable.Layout_ability_main);
7           startClockService();                    //启动 Service Ability
8       }
9
10      private void startClockService() {
11          Intent intent = new Intent();
12          Operation operation = new Intent.OperationBuilder()
13                  .withDeviceId("")
14                  .withBundleName(getBundleName())
15                  .withAbilityName(ClockServiceAbility.class.getName())
16                  .build();
17          intent.setOperation(operation);
18          startAbility(intent);
19      }
20
21      @Override
22      public void onActive() {
```

```
23          super.onActive();
24      }
25
26      @Override
27      public void onForeground(Intent intent) {
28          super.onForeground(intent);
29      }
30  }
```

至此,远程闹钟的完整示例就讲解完了。

7.5 本章小结

本章详细讲解了"远程闹钟"应用程序的开发过程,其内容涵盖前面章节讲解的大部分知识,包括基本界面控件的使用(如 Text、Button、Image、Checkbox、TimePicker、DatePicker 等)、高级界面控件的使用(如 ListContainer、CommonDialog 等)、Ability Slice 之间的跳转、事件处理(如单击事件、选择事件等)、关系数据库的使用等。本章还大致讲解了使用 Spring Boot 框架构建服务器。

通过本章的学习,读者将越来越熟悉 HarmonyOS 应用程序开发的一般步骤,逐步达到灵活运用所学知识的要求。

7.6 课后习题

1. 尝试在手机端应用中添加的闹钟模块中加入闹铃选择的功能,并且能在手表端播放选择的闹铃。

2. 尝试在手机端应用中加入通知功能,当手表端应用关闭闹铃后在手机端的通知栏处弹出一个通知,提示手机端用户闹铃已经被手表端用户关闭。

附录 A

SQL 语句使用简介

A.1 SQL 介绍

SQL 是 Structured Query Language(结构化查询语言)的简称,是一种标准的访问和处理数据库的操作语言。对于初学者来说,只要学会使用 SQL 对数据库进行数据插入、查询、更新和删除等操作即可。本附录将简要介绍 SQL 的使用方法,供读者在前述章节学习时参考。

A.2 SQL 项目表设计

本书举例的综合项目中,涉及闹钟时间以及闹钟状态等数据,可以把这些数据集中在一个 clock 表中显示,如图 A-1 所示。

clockid	years	months	days	hours	minutes	sends	happened
1	2022	5	1	8	0	0	0
2	2022	5	1	14	0	0	0
4	2022	5	2	14	0	0	0
5	2022	5	3	6	30	1	1
6	2022	5	3	14	0	1	1
7	2022	5	4	6	30	1	1
8	2022	5	4	14	0	1	1
9	2022	10	1	7	0	0	0
10	2022	10	1	14	30	1	1
11	2022	10	2	7	0	0	0
12	2022	10	2	14	30	0	0
13	2022	10	3	7	0	0	0
14	2022	10	3	14	30	0	0
15	2022	10	4	7	0	1	1
16	2022	10	4	14	30	1	0
17	2022	10	5	7	0	1	1
18	2022	10	5	14	30	1	1
19	2022	10	6	7	0	0	0
20	2022	10	6	14	30	0	0
21	2022	10	7	7	0	0	0
22	2022	10	7	14	30	0	0

图 A-1 clock 表

clock 表中的第一行代表各个属性,从左往右依次是闹钟的 clockid(闹钟编号)、years(年份)、months(月份)、days(日期)、hours(小时)、minutes(分钟)、sends(发送状态)、happened

(发生状态)。从上往下依次是各个闹钟的数据及设置情况。为了更加简洁地呈现表的属性结构,可设计一张属性结构表来记录属性结构数据,在该表中使用英文字段代替中文属性名称。针对该表的情况,由于属性都是整数,所以都使用 int(整数类型)作为数据类型。通常重要的属性值必须是非空的,否则会报错。最后,SQL 主键用于唯一地标识每一行,是该行数据相对于其他行数据的独一无二的标识符。表 A-1 给出了该表的结构。

表 A-1　clock 表结构

编号	名称	字段代码	类型	长度	是否可空	主键
1	编号	clockid	int	255	非空	是
2	年份	years	int	255	非空	否
3	月份	months	int	255	非空	否
4	日期	days	int	255	非空	否
5	小时	hours	int	255	非空	否
6	分钟	minutes	int	255	非空	否
7	发送状态	sends	int	255	可空	否
8	发生状态	happened	int	255	可空	否

A.3　创建 SQL 表

使用数据库管理、开发工具,可以便捷地操作 SQL 表及其数据。Navicat for MySQL 是一个功能齐备、图形化的数据库管理、开发和维护软件。图 A-2 为安装并配置好的 Navicat for MySQL 菜单界面,限于篇幅,其安装过程请读者参考有关资料。

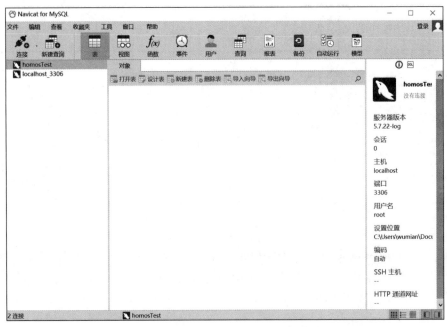

图 A-2　Navicat for MySQL 菜单界面

为了便于下文中学习 SQL 表格的操作，先在 Navicat for MySQL 中创建表 A-1 所指定的 clock 表，步骤如下。

（1）依次单击 Navicat for MySQL 左上角的"文件＞新建连接＞MySQL"，如图 A-3 所示。

（2）自定义输入连接名、用户名及密码，可以勾选"保存密码"便于下次创建，也可以单击"测试连接"按钮，查看是否连接成功，最后单击"确定"按钮完成创建，如图 A-4 所示。

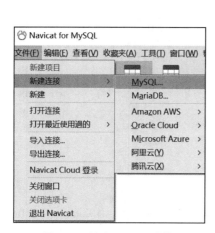

图 A-3　新建 MySQL 连接

图 A-4　创建 MySQL 连接

（3）双击 homosTest，若左边图标变绿，则表示连接成功，展开后可以对该连接下的数据库进行操作，如图 A-5 所示。

图 A-5　MySQL 连接成功

（4）右击 homosTest，从弹出的快捷菜单中选择"新建数据库"，使用 UTF-8 字符集以及 utf8_general_ci 排序规则，再单击"确定"按钮，如图 A-6 所示（UTF-8 是世界通用的语言编码，使得各国的文字都可以正常显示，不会乱码，utf8_general_ci 不区分大小写，校对速度快,但准确度稍差)。

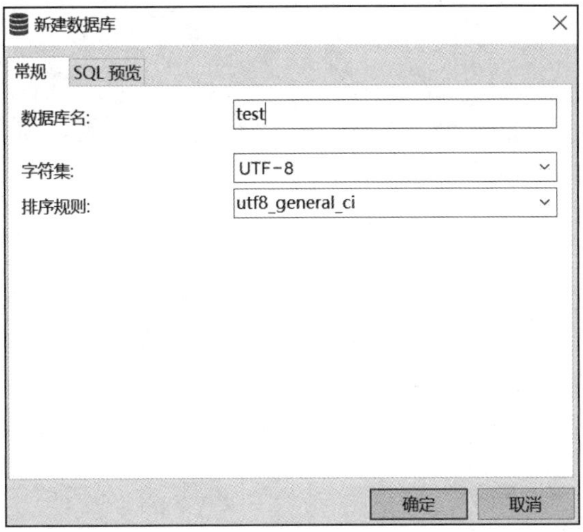

图 A-6　新建数据库

(5) 新建数据库后,双击 test 图标变绿后,表示数据库连接成功,展开后可以对该连接下的表或视图进行操作,如图 A-7 所示。

图 A-7　连接 test 数据库

(6) 单击"新建查询"图标,将两个下拉框分别调整为 homosTest 连接与 test 数据库,使用 SQL 语句对数据库进行操作,如图 A-8 所示。

(7) 按照表 A-1 使用 SQL 语句创建一个 clock 表,输入 SQL 语句后单击运行即可创建表,若信息框显示 OK,则表示表格创建成功,如图 A-9 所示。

在 clock 表中,由于各个数据都没有小数部分,所以都使用 int 类型。为了便于标识各个数据,使用 clockid 作为主键,并且人为规定 sends 为 1 时表示已发送,为 0 时表示未发送,happened 为 1 时表示已经发生,为 0 时表示未发生。除 sends 和 happened 属性可以为空,其他属性不能为空,否则操作时会报错。

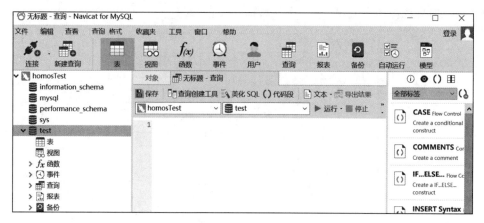

图 A-8　新建查询

（8）打开 test 数据库下的表，可以看到一个 clock 表，如图 A-10 所示（如果 clock 表没有显示，右击表进行刷新）。

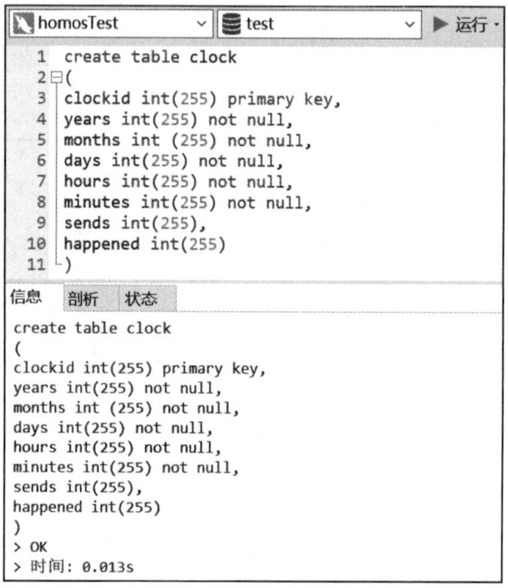

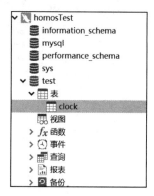

图 A-9　SQL 语句运行　　　　　　　　图 A-10　新建表格成功

（9）双击 clock 表后，由于还没对该表进行任何操作，因此它还是一个空表，可以手动向表内按格式自由输入数据，如图 A-11 所示。

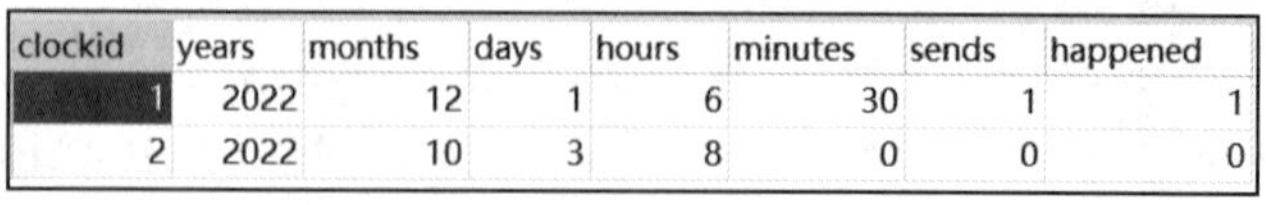

clockid	years	months	days	hours	minutes	sends	happened
1	2022	12	1	6	30	1	1
2	2022	10	3	8	0	0	0

图 A-11　clock 表结构

（10）为了方便学习，也可以直接将本教程的 SQL 文件通过 Navicat for MySQL 导入数据库中进行操作。首先双击激活数据库连接,如步骤(5)所示,之后右击数据库,从弹出的快捷菜单中选择运行 SQL 文件,再选择准备好的 SQL 文件,再单击"开始"按钮即可导入数据库,如图 A-12 所示。

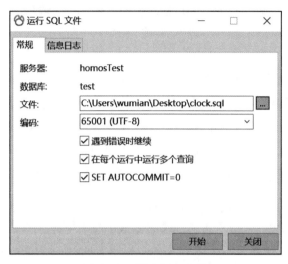

图 A-12 运行 SQL 文件

(11) SQL 文件运行成功后,打开 test 数据库中的 clock 表即可查看导入的表格数据,如图 A-13 所示。

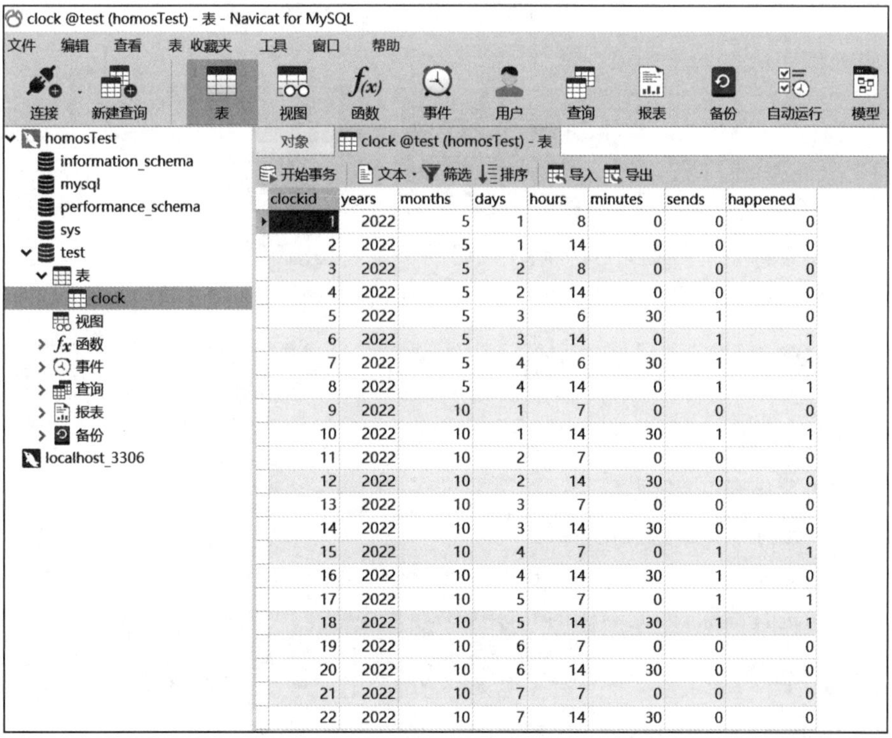

图 A-13 导入的表格数据

A.4　SQL 的查询、增加、修改、删除操作方法

前面介绍了用 Navicat for MySQL 创建表的操作，下面主要通过对图 A-11 所示的表格数据的操作实例了解 SQL 的查询、增加、修改、删除方法。

1. 查询数据——SELECT 语句

SELECT 语句的一般格式为：

```
SELECT    [ALL|DISTINCT]   <属性列1>,   <属性列2> …   [AS]   <别名>…
FROM    <表>
WHERE    <条件>
GROUP  BY   <列1>
ORDER  BY   <列2>   [ASC | DESC]
```

其功能是从 FROM 中的＜表＞中找出满足 WHERE 中的＜条件＞的各个＜属性列＞的数据，从而形成新的表格。AS 可以为表名称或列名称指定别名，让查询结果的列名称的可读性更强，还可以使用 DISTINCT 消除取值重复的行，并且可以使用 GROUP BY 将结果按＜列 1＞的值进行分组，该属性列值相等的元组为一个数据组。使用 ORDER BY 将结果按＜列 2＞的值升序（ASC）或降序（DESC）排列。

例 A.1　查询全部闹钟的详细记录。

在对数据库进行查询操作时，需要使用 select 命令从数据中查询数据，如果要将所有的记录从表中取出，则必须写出所有的属性列，这样才能完整地显示查询结果，因此本例具体代码如程序清单 A-1 所示。

程序清单 A-1

```
1  select clockid, years, months, days, hours, minutes, sends, happened from clock;
```

为了方便查询全部数据，便于 SQL 代码的书写，上述代码的属性列可以省略，同时使用 * 进行代替，最终效果一致，因此本例的具体代码如程序清单 A-2 所示。

程序清单 A-2

```
1  select * from clock;
```

查询结果如图 A-11 所示。

例 A.2　查询所有非重复的年份。

表中经常会有重复的列，为了省略重复的列，只显示其中一个，可以使用 select dinstinct 语句进行筛选，因此本例的具体代码如程序清单 A-3 所示。

程序清单 A-3

```
1  select distinct years from clock;
```

查询结果如图 A-14 所示,可知表中仅有 2022 年设置了闹钟。

例 A.3 查询所有发送但未发生的闹钟数据。

为了从表格中进一步过滤记录,实现更加细致的查询功能,可以在 where 子句后指定一个过滤条件实现此功能,也可以进一步使用 and 子句对数据增加第二个限定条件,进行更细致的过滤,本例要查询发送但尚未发生的数据,此前人为规定 sends 为 1 时表示闹钟已发送,为 0 时表示闹钟未发送,happened 为 1 时表示闹钟已发生,为 0 时表示闹钟未发生。可知本题 sends=1 并且 happened = 0,因此本例的具体代码如程序清单 A-4 所示。

图 A-14　dinstinct 查询结果

程序清单 A-4

```
1    select * from clock
2    where sends = 1 and happened = 0;
```

查询结果如图 A-15 所示。

图 A-15　where and 查询结果

例 A.4 查询所有非重复的闹钟月份数据并且按 clockid 降序排列。

此例与例 1 类似,但是需要按年份进行降序排列,可以使用 order by 关键字对结果集进行排序,order by 关键字默认按照升序(ASC)对记录进行排序,如果需要按照降序对记录进行排序,可以使用 DESC 关键字,因此本例的具体代码如程序清单 A-5 所示。

程序清单 A-5

```
1    select distinct months from clock
2    order by clockid DESC;
```

图 A-16　order by 查询结果

order by 查询结果如图 A-16 所示。

例 A.5 查询各年份的闹钟个数。

可以使用 count(属性列)方法查询指定条件的数据的个数,由例 A.5 题意可知需要使用 count(year)作为筛选条件,使用 count() 类似的方法时,可以搭配使用 group by 语句,根据列对结果进行分组。本例需要将闹钟个数与各个年份结合在一起,因此需要使用到 group by year 子句,同时,为了让查询结果的列名称的可读性更强,可以创建别名更形象地表示查询结果,所以本例使用 yearCount 作为列别名表示各年份的闹钟个数,因此本例的具体代码如程序清单 A-6 所示。

程序清单 A-6

```
1   select count(years) as yearClockCount from clock
2   group by years;
```

group by 查询结果如图 A-17 所示,可知 2022 年共设置了 22 个闹钟。

例 A.6　查询各个月份已发送并已发生闹钟数据的个数,并且按月份降序显示。

本例结合了之前几个例题的知识点,因此不做赘述。本例的具体代码如程序清单 A-7 所示。

程序清单 A-7

```
1   select count(clockid) as trueMonthClockCount
2   from clock
3   where sends = 1 and happened = 1
4   GROUP BY months
5   ORDER BY months DESC;
```

clockCount 查询结果如图 A-18 所示,可知 10 月有效闹钟数为 4 个,5 月有效闹钟数为 3 个。如果去除第三行 where 子句,该表数据将会变成 14 和 8,二者之和是例 A.5 的查询结果。

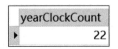

图 A-17　group by 查询结果

图 A-18　clockCount 查询结果

例 A.7　查询 2022 年 10 月 1 日至 10 月 7 日发送并发生的闹钟个数。

为了将查询条件限定在两个值之间的数据范围内,可以使用 between 操作符,具体格式为 where column_name between value1 and value2,因此本例的具体代码如程序清单 A-8 所示。

程序清单 A-8

```
1   select count(years) as dayClockCount
2   from clock
3   where days between 1 and 7
4   and years = 2022 and months = 10
5   and sends = 1 and happened = 1;
```

between 查询结果如图 A-19 所示,可知期间一共有 4 个闹钟发生并响应。

图 A-19　between 查询结果

2. 增加数据——INSERT 语句

INSERT 语句的一般格式为

```
INSERT  INTO  <表名>  [(<属性列 1>,  <属性列 2>…)]
VALUES  (<常量 1>,  <常量 2>…);
```

其功能是将新的数据插入指定表中，常量 1 的值对应属性列 1，常量 2 的值对应属性列 2，……以此类推。需要注意的是，如果在定义表时说明了 NOT NULLM，则不能取空值。

例 A.8　将新的闹钟数据插入 clock 表中。

本例中，插入新的闹钟数据时，需要注意某些数据必须是非空的，并且主键不能重复，否则会报错，同时数据格式必须与表的格式保持一致，因此本例的具体代码如程序清单 A-9 所示。

程序清单 A-9

```
1  insert into clock(clockid,years,months,days,hours,minutes,sends,
   happened)
2    values ('23','2022','12','31','8','30',1,0);
```

在 INTO 子句中指出了表名 clock，并指出了新增加的元组在哪些属性上赋值，同时不指定要插入数据的列名，只提供被插入的值也可以完成插入的操作。上述代码的属性列可以省略，最终效果一致，因此本例的具体代码如程序清单 A-10 所示。

程序清单 A-10

```
1  insert into clock
2    values ('23','2022','12','31','8','30',1,0);
```

插入结果如图 A-20 所示。

21	2022	10	7	7	0	0	0
22	2022	10	7	14	30	0	0
23	2022	12	31	8	30	1	0

图 A-20　插入结果

3. 修改数据——UPDATE 语句

UPDATE 语句的一般格式为

```
UPDATE  <表名>
SET  <列名> = <表达式>...
WHERE  <条件>;
```

其功能是修改表中满足 WHERE 子句条件的元组，使用 SET 将该元组<列名>的值改为<表达式>的值，WHERE 省略后所有的元组都会被修改。

例 A.9 将 clockid 为 5 的元组的 happened 属性修改为 1,表示闹钟已响。
本例的具体代码如程序清单 A-11 所示。

程序清单 A-11

```
1    update clock set happened = 1
2    where clockid = 5;
```

更新结果如图 A-21 所示,之前的效果可以参考附图 13。

图 A-21 更新结果

4. 删除数据——DELETE 语句

删除语句的一般格式为

```
DELETE
FROM    <表名>
WHERE   <条件>
```

DELETE 语句的功能是删除表中满足 WHERE 子句的元组数据。

例 A.10 删除 clockid=3 的闹钟数据。
本例的具体代码如程序清单 A-12 所示。

程序清单 A-12

```
1    delete from clock
2    where clockid = 3
```

删除结果如图 A-22 所示,之前的效果可以参考图 A-22。

图 A-22 删除结果

A.5 小结

附录 A 介绍了什么是 SQL,以及 Navicat for MySQL 的基本操作方法,其中包括建立数据库连接、创建 SQL 表、导入 SQL 文件等,并且通过代码的实际使用详细介绍了数据库中表的查询、增加、修改、删除功能。限于篇幅,SQL 的其他功能请读者按需自学,只要掌握以上的 SQL 知识以及操作技巧,就能开发本书的鸿蒙项目。

图书资源支持

感谢您一直以来对清华版图书的支持和爱护。为了配合本书的使用,本书提供配套的资源,有需求的读者请扫描下方的"书圈"微信公众号二维码,在图书专区下载,也可以拨打电话或发送电子邮件咨询。

如果您在使用本书的过程中遇到了什么问题,或者有相关图书出版计划,也请您发邮件告诉我们,以便我们更好地为您服务。

我们的联系方式:

清华大学出版社计算机与信息分社网站:https://www.shuimushuhui.com/

地　　址:北京市海淀区双清路学研大厦 A 座 714

邮　　编:100084

电　　话:010-83470236　010-83470237

客服邮箱:2301891038@qq.com

QQ:2301891038(请写明您的单位和姓名)

资源下载: 关注公众号"书圈"下载配套资源。

书圈

清华计算机学堂

观看课程直播